The First Darwinian Left

History and politics titles from New Clarion Press

Lawrence Black *et al.*, *Consensus or Coercion? The State, the People and Social Cohesion in Post-war Britain*

Duncan Hall, *'A Pleasant Change from Politics': Music and the British Labour Movement between the Wars*

Anne Kerr and Tom Shakespeare, *Genetic Politics: From Eugenics to Genome*

David Renton, *Classical Marxism: Socialist Theory and the Second International*

David Stack, *The First Darwinian Left: Socialism and Darwinism 1859–1914*

Leo Zeilig (ed.), *Class Struggle and Resistance in Africa*

Forthcoming

John Carter and Dave Morland, *Anti-Capitalist Britain*

Keith Flett and David Renton (eds), *New Approaches to Socialist History*

Mark O'Brien, *When Adam Delved and Eve Span: A History of the Peasants' Revolt of 1381*

The First Darwinian Left

Socialism and Darwinism 1859–1914

David Stack

New Clarion Press

First published 2003

New Clarion Press
5 Church Row, Gretton
Cheltenham GL54 5HG
England

New Clarion Press is a workers' co-operative.

A catalogue record for this book is available from the British Library.

ISBN paperback 1 873797 37 0
hardback 1 873797 38 9

Typeset in Times New Roman by Jean Wilson Typesetting, Coventry
Printed in Great Britain by The Cromwell Press, Trowbridge

Contents

Preface

It is a long established psychological principle that the language in which we think and communicate helps condition the thoughts we have and the conclusions we reach. That simple insight is the basis of my argument. Darwinism, broadly defined, provided the late-nineteenth-century left with an organic and an evolutionary language that profoundly influenced the form and content of its socialism.

I have not attempted to write a history of socialism, or even provide a comprehensive account of socialist thought in the period 1859–1914. My ambition was limited to providing an alternative narrative; one in which the various ways socialists grappled with a Darwinian inheritance – rather than class, continuity, betrayal or embourgeoisement – provide the connecting thread. There are, therefore, inevitably sins of omission and an occasional lack of balance. In mitigation I would plead that this is justified by the need to highlight the hitherto underestimated importance of Darwinism to the development of socialist thought and theory.

In a dual sense, this is a book of its time. With the apparent decline of class politics in western Europe, it is only to be expected that historians will reach into the past to construct non-class narratives around the birth and growth of the socialist movement. And at a time when Darwinism seems to be in renaissance – to such an extent that genetics professors are used to advertise motor cars – what could be more enticing than a narrative linking socialism with the most powerful science of the day? But this book is not a celebration of socialism's 'scientifity' and even less is it a paean to the influence of science in social thought.

This is a book I have written because I wanted to read it. Previous accounts had, I felt, failed to understand just how serious the first generation of socialists were about their science. The 'enormous condescension of posterity' still weighs heavily upon late-nineteenth-century socialists. Political historians – whose own scientific knowledge rarely extends beyond the ability to boil a kettle – continue to mock and condemn them for lacking a twenty-first-century understanding. But an even greater imperative than doing justice to one's ancestors motivated me. I wanted to help forestall a growing movement to use Darwinism as the foundation for a new politics of the left. I hope my account of the 'first Darwinian left' provides some measure of warning about the folly of such a course of action.

The academic inspiration for this book came from the staff and students of Queen Mary College, London, especially Daniel Pick and Maria Sophia Quine, who by inviting me to teach on their 'After Darwin' course triggered an interest in this subject, and Sarah Tasker, whose constant questioning forced me to clarify my own thoughts and conclusions.

Especial thanks are due to Annie Mitchell, for her help with the final stages of the manuscript, and to my family and friends who supported me through a difficult time in 2001.

David Stack
February 2003

Introduction: Myths and Misunderstandings

There is a widespread and persistent myth about the relationship between Darwin and the left. The myth is that Marx offered to dedicate some volume or edition of *Capital* to Darwin. A profusion of scholarly articles since the mid-1970s have failed to shift this illusion from popular perceptions. The documentary evidence was always slender. In the 1890s a letter to Edward Aveling – the partner of Marx's daughter Eleanor – got mixed in with a box of Marx's correspondence that was in Eleanor's care. The letter was from Darwin, politely declining Aveling's offer to dedicate an atheistic pamphlet to him. To one unwitting 1930s Moscow archivist it appeared that Darwin was refusing Marx and, in the hands of the historical profession, a legend was born.[1]

That legend flourished in a fertile soil. In the late 1890s and early 1900s in particular, a welter of literature had been produced expounding the intellectual compatibility of Marxism and Darwinism. Aveling himself had penned an enthusiastic piece for the *New Century Review* in 1897 entitled 'Charles Darwin and Karl Marx: a comparison', which was subsequently reprinted as a pamphlet and translated for distribution in a number of languages.[2] Enrico Ferri's *Socialism and Positive Science* – an ambitious attempt to unite the doctrines of Darwin and Marx with those of Herbert Spencer – had already appeared in Italy in 1894. And in Germany, Ludwig Woltmann's *Darwinian Theory and Socialism* (1899) and Karl Kautsky's *Ethics and Historical Materialism* (1906) headed a host of publications linking Marxism and Darwinism. With the intellectual affinity apparently proved, it seemed perfectly natural that a letter should be found confirming Marx's personal regard for Darwin.

Much of the blame for this myth must also rest with Engels. The turn of the century literature had taken its cue from Engels' graveside eulogy to Marx – 'Just as Darwin discovered the law of development of organic nature, so Marx discovered the law of development of human history.' Nor was this remark, reproduced in *Der Sozialdemokrat* for the digestion of socialists all over Europe, an aberration. In the preface to the fortieth anniversary edition of the *Communist Manifesto*, and a number of his other later works, Engels explicitly encouraged socialists to regard Marx and Darwin as complementary.[3] But significantly, and unlike some of his successors, Engels never sought to probe too deeply *how*. His purpose was purely polemical. Engels was not only keen for Marxism to bask in the reflected glory of Darwinism and prevent the new science being erected into a further barrier against socialism. Perhaps more pertinently, he was also desperate to beat off putative socialist rivals in Germany, such as Ludwig Büchner and Friedrich A. Lange, who as early as the 1860s and 1870s had begun to associate Darwinism positively with their alternative brands of socialism.[4]

A German debate over the compatibility of socialism and Darwinism ensued, provoking a public spat between the country's leading biologists in 1877 and causing the usually apolitical Darwin to comment upon the 'foolish idea [that] seems to prevail in Germany on the connection between Socialism and Evolution through Natural Selection'.[5] Engels' remarks are therefore best understood both as part of a general polemic to stave off the use of Darwinism as an anti-socialist instrument and as part of a more parochial propagandist campaign to steal the Darwinian mantle, and any kudos that went with it, from rival socialists. It was a tactic of which Marx fully approved and which he initially helped orchestrate.[6] However deep and genuine their regard for Darwin, Marx and Engels sought to associate their socialism with his science for purely tactical reasons. Marxism stood alone, philosophically distinct from Darwinism.

But what was true of Marx and Engels was not, as is sometimes assumed, true of the rest of the late-nineteenth-century left. Marx and Engels were the exception, not the rule. For the mainstream left in late-nineteenth-century Britain and Europe, Darwinism was integral to socialism. On one level this was a simple matter of chronology. Marx and Engels had a fully developed and coherent intellectual system prior to the publication of Darwin's *Origin of Species* (1859). They were of the last generation of socialists who could claim this. Thereafter socialism, by necessity and choice, was developed within a Darwinian discursive space. An accident of timing ensured that there was little by the way of conscious choice involved for most socialists. By the time the *Origin* appeared in 1859, older varieties of socialism, principally Owenite and Saint-Simonian, had largely died out. The socialism and socialist movements that arose in the next half-century were forged and matured in an era when Darwinism was an established part of the 'mental furniture'. The organizations and personnel of late-nineteenth-century European socialism were post-1859 products. In Britain it was only a quarter of a century after the *Origin* that the first socialist organizations of any lasting significance came into being, and not until 1893 that the Independent Labour Party (ILP) was formed. On the continent it was a similar story. Although the German Social Democratic Party (SPD) was founded in 1875, the French Socialist Party was only finally unified in 1905. And all the major late European socialist parties were most definitely *of* the late nineteenth century and, therefore, deeply hued with Darwinism.

The connection between socialism and Darwinism went deeper than a mere coincidence of timing, although to read most political historians one would think that socialists existed vacuum-wrapped before 1859 and, untouched by the Darwinian revolution, emerged only later to consider its consequences for their hermetically sealed politics. In fact, most of the socialists we will study were Darwinists before they found their political orientation. A biological language and way of thinking often preceded their politics. This ensured that the relationship between socialism and Darwinism was not purely instrumental. Darwinism was not simply a useful tool to be picked up or discarded at will. Nor was it simply a convenient cover for an independently constructed political argument. Darwinism was woven into the pattern of late-nineteenth- and early-twentieth-century socialism. The two were so intertwined that it makes no sense to regard one as prior to, or making instrumental use of, the other. The language of Darwinism became, for a time, the language of socialism. This mattered profoundly. A biological lexicon provided the structure and boundaries for political and philosophical discourse, and socialists developed their politics amidst evolutionary precepts and in an organic language. Darwinism provided socialism with its constitutive metaphor: a discursive space and a heuristic stimulus for the construction of a new version of socialism that was distinguishable from both pre-Darwinian Marxism on the left and liberalism on the right. It is these versions, found in the ILP in Britain and the Revisionist movement on the Continent, which constitute the first Darwinian left.

The new Darwinian left

The rediscovery of the first Darwinian left has been made all the more urgent by the threat of a second incarnation. That there should be a movement to create a contemporary Darwinian left ought to come as no surprise. Much like the era of the first Darwinian left, we are once again living in an 'age of biology' where Darwinism is feted as 'the single best idea anyone has ever had', and used to explain almost every conceivable element of human behaviour.[7] It was probably only a matter of time before someone was found to argue the need for the left to base its politics on Darwinism. Step forward Peter Singer, whose book *The Darwinian Left* (1999) was published as part of the 'Darwinism Today' series. The last detail is significant because the series was sponsored by Darwin@LSE. This group, fronted by the disarmingly plausible Helena Cronin, are not quite the straightforward Darwinists they would have us believe. They are, rather, evolutionary psychologists – a hybrid species, produced from a crossbreeding of evolutionary biology and cognitive psychology.

The extent to which Singer is bound up in their agenda can be seen in his suggestion that men work with disabled children in order to increase their attractiveness to potential sexual partners.[8] The absurdity of this proposition can be explained only by the propensity of evolutionary psychologists to relate as much human behaviour as possible to the sex-drive. In this sense the science is like Freud, except in place of childhood traumas the evolutionary psychologist seeks to recover the sexual urges of Stone Age man. In particular, evolutionary psychologists have been virulent in asserting that men and women are programmed for different patterns of sexual behaviour, with men promiscuously seeking to spread their genes indiscriminately and women seeking more stable relationships. This can lead either to innocuous nonsense, such as one evolutionary psychologist seeking an evolutionary explanation for why he wants to have sex with young women, or to the dangerous assertion that rape is an evolved male 'mating strategy'.[9] Singer's association with this group is relevant because it cuts to the heart of what is wrong with his analysis – philosophically, politically and historically.

Philosophically, there are two weaknesses. The first is that Singer's argument that 'it is time to develop a Darwinian left' rests on the rather naïve positivistic presumption that there are straightforward and incontestable lessons to be learnt from science.[10] This is maintained by a deft sleight of hand that confounds evolutionary psychology with Darwinism. Without this manoeuvre, Singer would be forced to concede that Darwinism is not a sealed and settled set of politically neutral 'truths', in front of which all politics must bow. Indeed, one of the most contentious areas within Darwinism is evolutionary psychology. Many leading Darwinists reject it out of hand. The recent attempt by Steve Jones, for example, to 'update' Darwin's *Origin* in *Almost like a Whale* (1999) passed over the discipline in an embarrassed silence, whilst the evolutionary biologist Steven Rose is an outspoken opponent. But if Singer had admitted that Darwinism is not an unproblematic and uncontested body of knowledge from which the left can draw simple and indisputable lessons, then his prescriptions for the left would have carried a lot less force.

Singer's background as a philosopher of animal rights might explain this 'blind spot' for evolutionary psychology, which leads him into his second major philosophical weakness. The attraction of evolutionary psychology, for those interested in animal welfare, is that it narrows the gap between humans and 'non-human animals'. But instead of doing this by elevating the position of animals, it works in the opposite direction by reducing humans to a set of bestial impulses and desires. The result is an unattractive view of both animals and men, and a flagrant disregard for the influence of 7,000 years of culture in shaping human behaviour. Such a mode of reasoning slips all too easily into a politically conservative naturalization of existing social practices and arrangements, and leaves Singer explaining the lack of female chief executives in terms of women's attitudes to work and reproduction.[11]

The pre-Darwinian left, Singer argues, entertained impossibilist dreams about perfectibility, ending conflict and eradicating inequality, through tackling discrimination, prejudice, oppression and poor social conditions. Singer's Darwinian left, by contrast, is far more sanguine and recognizes that human nature is responsible for many of these conditions. Politically, therefore, Singer offers a 'sharply deflated' vision of the left.[12] And by focusing on what is static and unchanging in human nature, rather than what is cultural and therefore malleable, Singer severely limits the scope for any programme of reform. This chimes well with third way politics but suffers from the same inadequacies as that political brand. By his own admission, Singer has no answers to the challenges of globalization and the World Trade Organization.[13] He has no understanding of economic or power relationships in society and no means to analyse them. Worst of all, he can provide little solace for 'the weak, poor and oppressed', bar some well-intentioned platitudes.[14]

What makes Singer's argument dangerous is that he gives the third way an added veneer of scientific authority. This is deployed to short-circuit moral and political argument by an appeal to the unchallengeable authority of Darwinism. Singer even goes so far as to deny the validity of any argument that is not rooted in biology. Thus Locke's rebuttal of Filmer is deemed inadequate because it was not made on an evolutionary basis.[15] This search for scientific certainty may be understandable in our postmodern age but should be fiercely resisted by all those interested in developing a truly radical left. Although he denies decoding the 'ought' from the 'is', Singer's argument leads him into naturalizing existing social relationships and providing a biological apologia for sexism, racism and inequality.[16]

Historically, Singer's account is both ignorant and misleading. Third way politics is notoriously lacking in historical understanding and Singer's book is no exception. The position of the so-called old left is hideously caricatured in order to justify the ditching of a socialist agenda. The whole analysis rests upon asserting that Marx denied the notion of human nature and that Marx's position was that of the rest of the nineteenth- and twentieth-century left.[17] We shall see in Chapter 5 that Marx's notion of human nature was, in fact, far more complex than Singer allows and throughout we will argue that Marx did not dominate left-wing thinking in the way Singer seems to believe. Nor are these the only errors. The force of Singer's argument is that he knows better because he recognizes the truth of Darwinism. Every single preceding left-wing thinker, with the possible exception of Kropotkin, is nonchalantly dismissed because they did not share this insight.

But, as we shall see, Singer is not the pioneer he would portray. He excuses the failings of his late-nineteenth-century forebears by blaming 'the limitations of Darwinian thinking in that period', but the real limitations are in Singer's historical knowledge.[18] The late nineteenth century witnessed a wonderful flowering of Darwinian thinking, in which the cross-fertilization of socialism and Darwinism was initially a much stronger strand than the association of that science with the political right. Of course, to concede this would plunge Singer's argument into disarray. His straw man caricature of a resolutely pre-Darwinian left would unravel and he would be drawn to consider those who have preceded him in proclaiming a Darwinian left. But Singer may not have been wilfully ignorant. The historical profession must bear a burden of responsibility for its failure to take seriously the role of science in the development of socialist thought.

Two permeable systems

Annie Besant (1847–1933) could not have been more explicit: '*I am a Socialist because I am a believer in Evolution.*'[19] This epigram, the first justification offered in her personal manifesto of 1886, *Why I am a Socialist*, was to be repeated, though with less pithy directness, by countless other socialists. Yet the connection, and more particularly the order of precedence, is rarely recognized by political historians. Our primary task is to alert the reader to the importance of the relationship between Darwinism and socialism in the period 1859 to 1914. Along the way we hope to persuade political historians to think about this relationship in a different way. We need to take seriously the left's engagement with science and understand its positive contribution to the development of socialism. Instead of regarding the role of science in political thought in purely instrumental terms – as providing a justification for a prior political position – we need to recognize that there was a genuine isomorphic interaction of socialism and Darwinism in the late nineteenth century. It was not a case of a discrete and established political theory self-consciously adapting to an equally discrete and established scientific theory; or of socialists picking and choosing convenient 'scientific' ideas and terms, whilst remaining fundamentally uninfluenced by the tools they acquired. Neither Darwinism nor socialism was a fully developed system of thought. Both were in their infancy and both, therefore, were permeable, adapting and evolving frameworks. Once this is recognized, the relationship that existed between the two 'isms' becomes far more interesting and significant than is usually allowed.

Those studying the pre-Darwinian period have already begun to fill the 'veritable wasteland' that has for many years constituted the scholarly research into the role of science in the history of the left.[20] It remains for those working in the Darwinian period to do the same. For many years the use of science by Owenites, radicals and Chartists was regarded with suspicion. When not betraying an 'unashamedly instrumentalist' attitude, it has been taken to signal a fatal embrace of a corrupting bourgeois politics.[21] The organic sciences of the nineteenth century were seen to represent a 'momentous departure' in the movement 'from equality to organicism'.[22] But much recent work on Lamarckism, biology, physiology and phrenology has challenged this attitude. It has not only demonstrated the centrality of the organic sciences to Owenite, radical and Chartist thought, but has also shown that this was an isomorphic, rather than an instrumental, relationship with non-fatal consequences.[23] Clearly this fruitful interaction did not come to an abrupt end in 1859. Quite the opposite. Darwin's *Origin* furthered encouraged the left to engage with science and more closely bound the social and the scientific together by proposing 'one general law, leading to the advancement of all organic beings'.[24]

This was made possible by the fact that Darwinism was not a settled system. In the late nineteenth century, Darwinism was even more highly contested than it is today. Not even Darwin was certain about the details of his doctrine. The Darwin of the sixth, heavily revised edition of the *Origin* was, for example, far more Lamarckian than the Darwin of the first edition. There is, therefore, no need to mock or dismiss out of hand those socialists whose ideas of Darwinism do not match up to our modern understanding. The narrow definition of Darwinism as evolution by natural selection, and nothing more, did not emerge sealed and sacred from the pages of the first edition of the *Origin*. It was itself *constructed* in the late nineteenth century by the new breed of professional scientists clustered around T. H. Huxley, and was only decisively confirmed by the 'modern synthesis' of the inter-war period.[25] Prior to this, Darwinism had been a constantly shifting, ill-defined complex of contesting definitions, in which a socialist exegesis of the *Origin* played its part.

Instead of anachronistically looking for a pure Darwinian standard, which the socialists inevitably failed to meet, it makes more sense to talk of Darwin*isms* in the late nineteenth century than it does of Darwin*ism*. A positivistic framework that treats Darwinism as a hermetically sealed doctrine, to be accepted or rejected *in toto*, is an outdated way in which to think about any scientific theory. Much of the most interesting recent work in the history of science has emphasized how Darwinism was a 'situated science', differing in content and meaning according to the demands of its differing audiences in Britain, Europe and America.[26] And this approach could be applied to different political as well as national audiences. Just as Darwinism in Russia was a far more co-operative doctrine of nature than the rabidly individualistic and competitive version presented in the USA, so it was possible to construct a socialist Darwinism.

Indeed, this was not only possible, but absolutely necessary – not only because Darwinism was the dominant idea of the age, but also because the notion of separate spheres of politics and science had not yet been established. Both pre- and post-1859 there was an inextricable intertwining of social and scientific theories. This is why the term 'social Darwinism' is of questionable value and will be eschewed in this study. It neatly captures the spirit of applying a biological methodology and framework to social problems and questions but is misleading to the extent it suggests that there was, or could be, a 'non-social Darwinism' of pure, unadulterated science. There was no Darwinism prior to, or independent of, social Darwinism. They emerged as one from the 'common context' of nineteenth-century social and scientific thought that pre-dated the *Origin* and contributed to an evolutionary and organic discourse that encompassed both scientific and social debate.[27] Moreover, the term 'social Darwinism', as originally understood by Hofstadter, indicated an individualistic interpretation, which only later gave way to a collectivist exegesis.[28] Our argument, however, is that the left were engaging with Darwinism from its earliest formulation, and that the language and heuristic structures of Darwinism contributed to the development of socialist thought. The concepts, terminology and assumptions present in Darwinism stimulated the construction of analogous structures that directed and conditioned late-nineteenth- and early-twentieth-century socialism.

This argument has implications for the long-running historiographical dispute about the extent of 'continuity' on the left and, more particularly, the nature of the transition from radicalism to socialism. Nowhere amidst all the wind and hot air expended in dowsing for the 'currents of radicalism' flowing into the early Labour Party has there been any consideration of the importance of developments in science as a factor in the recasting of left-wing thought. The 'currents' that Alastair Reid and Eugenio Biagini identified have been strictly political and programmatic.[29] This is a mistake. As some heretical voices within the temple have pronounced, the same political programme often developed a new meaning in a changed political context and the convergence of policy positions often obscured deeper philosophical divides.[30] There is, therefore, a need to probe below the superficial continuities between radicalism and socialism and to examine 'the "intellective" elements' underlying left-wing politics.[31]

One of these elements was science. And at the very moment that radicalism was passing into socialism, the reception and diffusion of the Darwinian revolution – so fundamental that it threatened to alter basic conceptions of the universe and man's place within it – was taking place. This raises the hitherto unexplored question of whether or not after Darwin the left made the same programmatic arguments on an entirely new philosophical foundation. After all, the basic premise of radicalism – the providential assumption of a beneficent nature – was dealt a blow by Darwin every bit as deadly as that aimed at Paleyean design. And it is our contention that Darwinism was a key factor in the movement from radicalism to socialism. The change may not have been sharp or caesural, but Besant's epigrammatic explanation of her own political odyssey contains a wider truth. The intellectual readjustment necessitated by Darwinism, and more particularly an exploration of the questions Darwin left unanswered, set the left on a path that led from radicalism to socialism.

1

Darwin's Challenge

The publication of Darwin's *Origin of Species* in 1859 set the left a challenge: how to reconcile a philosophical belief in equality, co-operation and education with a scientific doctrine of inequality, competition and inheritance. But only in retrospect can this challenge be seen with such clarity. Few, if any, contemporaries identified an antagonism or felt a dilemma. The book's initial reception on the left ranged from indifference to enthusiastic embrace. John Stuart Mill, the leading liberal philosopher of the day, simply ignored Darwin – much to the latter's chagrin.[1] The old English radical Thomas Hodgskin similarly failed to mention Darwin or the *Origin* in the weekly science column he wrote throughout the 1860s.[2] Marx, by contrast, had re-read the *Origin* twice by 1862 and wrote gushingly to Engels that the book 'contains the basis in natural history for our view'.[3]

Marx was soon to temper his enthusiasm, but it is significant that while liberals and radicals paid little heed to the new science, socialists were keen to find in Darwin a confirmation of their politics. They were encouraged by the fact that up until 1859 the most advanced science and the most advanced political positions had seemed to march together, hand in hand. In particular, early socialists had drawn inspiration from the evolutionary theory of Jean-Baptiste Lamarck (1744–1828), the most influential of the scientific innovators to emerge in post-revolutionary Paris.[4] Darwinism, however, was to prove a more double-edged sword. Even as it confirmed the fact of evolution, it did so by proposing a mechanism – natural selection – which seemed to undermine the left's most cherished ideals.

In fact, what Darwin bequeathed to the left was a 'curate's egg' of contradictions. The great boon of the *Origin* was that it decisively confirmed the 'fact' of evolution; the drawback was that it did this by enshrining the Malthusian struggle for existence in nature. But below the surface of such an apparently sharply defined dichotomy, the *Origin* posed a more complex set of unresolved dilemmas. On a host of issues Darwin was unclear and the *Origin* offered mixed messages that were neither necessarily positive nor negative for the left. In this category fell such important questions as the relative importance of inheritance and environment, the relationship between evolution and progress, the choice of atheism or spiritualism, and the balance of individualism and collectivism in natural selection.

For the next fifty years socialists contributed to articulating, probing and attempting to resolve Darwin's unsettled questions. It mattered greatly to a generation of socialists raised upon Comtean positivism that their answers were compatible with the most advanced scientific position, but there was no settled tribune of scientific 'truth' that they could consult. Darwinism, as much as socialism, was 'under construction'. And as both grappled with the problems Darwin had posed, they inevitably drew upon many of the same raw materials and incorporated many of the same assumptions and structures into their systems of thought. This is not to argue that socialists and Darwinists always gave the same answers, merely to point out that they both developed through addressing the same set of questions. As they did so, socialism probably became less environmentalist and more hereditarian, just as Darwinism became less individualistic and more collectivist. Neither resolved its attitude to progress or religion. Both found teleological assumptions difficult to eradicate and tactfully both left God's lingering death unannounced. The year 1859, therefore, signalled neither a conservative counter-revolution nor a positive scientific endorsement of socialist politics, but the posing of a set of questions that were profoundly to influence the future development of the left.

Evolution before Darwin

Prior to 1859 the concept of evolution was firmly associated with the radical and socialist left in Britain for both philosophical and historical reasons. Philosophically, by dissolving all rigidity in nature and suggesting that everything in the universe was contingent and transitory, rather than eternal, the concept of evolution had inevitably assumed a central place in the armoury of social critics. An image of nature in constant flux, as opposed to the Christian view of a static creation, opened up the possibility that society would demonstrate a similar necessity for change; that not only could things be otherwise, but they should and would be so.

This philosophical alliance had been secured in the French revolutionary debates of the 1790s. Charles Darwin's grandfather, Erasmus, penned a cod-poetic work entitled *Zoonomia* (1794–6), which offered an organic and evolutionary vision of nature and drew a virulent condemnation from the conservative Anglican theologian William Paley.[5] Paley's famous 'watchmaker analogy' of a mechanistic universe formed the basis of conservative thought in a period of anti-French reaction and was integral both to the politically complacent assertion that 'Whatever is, is right' and the *Reasons for Contentment* that Paley offered to the poor.[6] Thereafter, evolutionary notions suffered from politically subversive and foreign associations, which were confirmed by the subsequent development of evolutionary biology in post-revolutionary Paris. With Lamarck's theories untaught in the British universities, in the 1830s and 1840s it fell to socialist journals such as *The Oracle of Reason* to propagate the new evolutionary science.[7]

Nonetheless, even before Darwin published the *Origin*, a counter-movement was under way to make evolutionary ideas less politically dangerous. The key text was an anonymous work entitled *Vestiges of the Natural History of Creation* (1844). Written by the dictionary publisher Robert Chambers, and bringing together evolutionary accounts of astronomy, natural history, geology, chemistry, physics, phrenology, political economy and anthropology, the *Vestiges* presented evolution as the law of all nature. Although later neglected, the *Vestiges* played an important role in bridging the gap between the radical Lamarckian approach of *The Oracle of Reason* and Darwin's theory. The work was tremendously successful, with eleven editions selling over 23,000 copies between 1844 and 1860.[8] Chambers' great contribution lay in imitating the success of his hero, another anonymous author, Sir Walter Scott. For just as Scott's *Waverley* series had turned the novel from an instrument of debauchery into one of self-improvement, so the exciting journalistic style of the *Vestiges*, and the providential veneer Chambers gave to evolutionary ideas, brought another previously dangerous genre into the middle-class parlour.[9] Chambers' success in this venture partly explains the lack of hostility that greeted the publication of the *Origin*. The public had been long prepared to accept evolutionism. Chambers had gently roughed up Paley's static and mechanistic world view before Darwin delivered the knock-out blow.

Darwin's key innovation was to change the mechanism through which evolution was said to work. Chambers had been unconvincing on this. In the *Vestiges*, evolution was an embryological unfolding of a pre-ordained plan, the realization of a Providential design. There was no real attempt to identify any mechanism through which change occurred. Lamarck was a more serious biologist. He had attempted to explain the evolutionary process – the variation and progressive diversification of plant and animal life from its ancestors – through the inheritance of acquired characteristics. Explained simply, this meant that any physical changes achieved in one generation – through use or disuse, conscious effort or a response to the environment – were passed on to the next. Lamarck's was a benign and straightforward account. Darwin, by contrast, explained evolution in terms of a harsher process of natural selection.

The difference between Darwin's and Lamarck's doctrines is easily illustrated by their respective answers to the hackneyed question that Lamarck attempted to answer, 'Why do giraffes have such long necks?' An inheritance of acquired characteristics suggests a benign process of stretching and incremental improvement. The use of the neck-muscles, an environment of tall trees and a conscious effort to stretch, all contribute to a parent passing on to their offspring an increased neck-length. Natural selection, by contrast, suggests that the giraffes' neck-length is the outcome of a brutal struggle for life in which shorter-necked ancestors are wiped out. According to Darwin, all plant and animal species had a tendency to overpopulate their potential food resources. This induced an intra-species struggle for life in which the 'fittest' or 'best adapted' survived and the 'unfit' or 'poorly adapted' were eliminated. In natural selection the 'survival of the fittest' was the root of evolutionary change, and utility in securing scarce resources led to the inheritance of certain traits. In the case of the giraffe, a longer neck offered access to higher leaves and, therefore, more food. This ensured a higher survival rate among longer-necked animals and therefore a greater success in breeding, which in turn produced a progressively longer-necked species. Very different political imperatives flowed from these differing accounts of evolutionary change.

The politics of Darwinism

Writing in 1877, Darwin described his politics as 'Liberal or Radical', but his overriding concern was to project an 'apolitical' image.[10] With the exception of joining Mill's Jamaica Committee, and despite occasionally letting his guard slip in personal correspondence, Darwin studiously avoided political pronouncements and activity.[11] It has taken an outstanding work of biography by Adrian Desmond and James Moore to recapture the deeply political nature of Darwin's work. Inspired by Robert Young's war cry that 'science is social', Desmond and Moore set out to 'unpack' the political and ideological contexts in which Darwinism was forged. In doing so they shattered the carefully preserved illusion of the saintly, apolitical Darwin and rediscovered the roots of evolution by natural selection in the Malthusian Poor Law culture of 1830s and 1840s Britain. Darwinism, they argued, was a product of the atomistic and individualistic society that Thomas Carlyle railed against; of the utilitarian values that Charles Dickens satirized; and of the 'weak to the wall', 'devil take the hindmost' ethics that the late twentieth century lauded as 'Victorian values'. In this Desmond and Moore have performed an invaluable service but their work leaves unresolved the nature of the challenge that Darwin posed for the left.

Two different approaches might be deduced from their work. The first would be to accept the identification of the Malthusian, utilitarian and *laissez-faire* elements in the formation of Darwinism as foreclosing any meaningful discussion of how that theory was later received and utilized. The context in which Darwinism was forged, that is, might be held to condition and limit its subsequent reception and use. This certainly seems to be the view of Desmond, who in other works has described the publication of the *Origin* as a 'counter-revolution'.[12] It is also a view that accords with the stance taken by many late-nineteenth-century critics of socialism. For example, in 1877, the German zoologist Ernst Haeckel angrily dismissed any connection between socialism and Darwinism as 'about as compatible as fire and water'. Darwin's theory of evolution, Haeckel argued, implied an ever-increasing differentiation among men, which rendered socialists' egalitarian beliefs 'a fathomless absurdity'. Moreover, the 'cruel and merciless struggle for existence which . . . in the course of nature *must* rage' doomed the majority to perish in competition, with only the 'fittest' to survive. Natural selection ensured an 'aristocratic' not a democratic outcome, and was therefore fundamentally and necessarily antagonistic to socialism.[13] Darwin himself reached a similar conclusion about the conservative implications of his theory in his personal correspondence in 1879. But this was fully twenty years after the publication of the *Origin* and contrasts sharply with the pre-publication doubts, fears and wracked nerves that long prevented Darwin going public with a theory he had fully worked out in his *Notebooks* of the late 1830s.[14]

It is in Desmond and Moore's tale of the twenty-year travails that Darwin endured – as he repeatedly postponed publication of a theory that at times felt 'like confessing a murder' – that we can identify an alternative, less deterministic, approach to understanding the challenge Darwin posed.[15] Taken with the perennial resurfacing of his theory's relationship to socialism – Haeckel's retort simply encouraged socialists to explore the link – Darwin's own doubts highlight the profound political ambiguity inherent in Darwinism. This ambiguity has been obscured by an overemphasis on the political centrality of Malthusianism and an insufficient appreciation of the indeterminacies in crucial areas of Darwin's thought. Certainly Darwin's reading of Malthus' *Essay on the Principle of Population* (1805) was the crucial catalyst in the development of his thought and his description of the operation of evolution by natural selection in the animal and vegetable kingdoms was 'the doctrine of Malthus . . . with tenfold force'.[16] But just because Malthus was the most obviously 'non-scientific' inspiration for Darwin, it does not mean that we should assume Malthusianism controlled the political character of Darwinism. Indeed, in many ways the openly ideological character of Malthusianism made it more susceptible to criticism than other more 'scientific' influences.[17] For many socialists 'Darwin the transmutationist' was a stronger image than 'Darwin the Malthusian'. The Darwinian concoction was an unstable mix of ingredients. As the cauldron from which the 'modern synthesis'[18] was eventually to be served simmered throughout the late nineteenth and early twentieth centuries, socialists and non-socialists alike added to it and tried to extract their favourite flavours. In the process the unappetizing Malthusianism strain periodically assumed greater or lesser strength, as other more tempting tastes vied for dominance.

Moreover, there has been an insufficient sensitivity to the differences between socialism and radicalism. Just because Malthus was the *bête noire* of radicalism did not make socialism implacably hostile to all things Malthusian. Although socialists –Marx and Engels included – inherited from radicalism a residual hostility to Malthus and his *Essay*, philosophically socialists were able to take a far more relaxed attitude.[19] To the extent that socialism was both less concerned to demonstrate an inherent harmony in nature – arguing for the human ability to construct the new utopia in a cultural space – and correspondingly more inclined to embrace birth control, it was able to broach a compromise with Malthusianism. This process, which had begun even in the early socialism of the Owenites – many of whom were atheists and endorsed birth control – was pursued with vigour by Besant and others in the 1870s and 1880s.[20] Few socialists endorsed John Robertson's call to claim Malthusianism for the left,[21] but many agreed with Engels' view that even if Darwinism rested upon 'the crudest and most barbarous theory that ever existed', it still had much to offer.[22]

This was important because it enabled socialists to embrace Darwinism without worrying too much about the Malthusian spectre, and it goes some way to explaining why socialism was able to usurp radicalism as the defining ideology of the left. Socialists were aided in this task by a changing interpretation of Malthusianism itself. By the mid to late nineteenth century the Anglican conservative philosophy of 'Parson Malthus' had been improbably reinterpreted into a secularist ideology, which gave its name to an organization campaigning for birth control. Few actually read Malthus and most were likely to know his name through the activities of Charles Bradlaugh and Annie Besant's Malthusian League. More fundamentally, socialism did not share the deistic and providential view of nature upon which radicalism was premised and which philosophically defined its very core in opposition to Malthusianism.[23] Thus on closer inspection the notion of an immediate and necessary antagonism between socialism and Darwinism, rooted in Malthusianism, is overstated.

Darwinism's unresolved issues

And what was true of Malthusianism was also the case in four other areas in which Darwinism has been seen as inaugurating a decisive move against the left.[24] The first of these was the question of inheritance. It is easy to assume a necessary antagonism between a rigidly hereditarian Darwinism and an environmentalist and Lamarckian left. In fact the lines of division were nowhere near as distinct. Darwinism may have shifted the balance towards hereditary limitations but it was far from settling where the boundary between nature and nurture should be drawn. The 'hard' heredity theories of genetics, which are now called Darwinian, did not triumph until the mid-twentieth century. Darwin's own theory of inheritance – pangenesis – was a much 'softer' theory, easily acceptable to socialists, and quite perversely the *Origin* sparked a renewed interest in Lamarckism.[25] Successive editions gave progressively greater importance to explanations based upon use and disuse, the environment and the inheritance of acquired characters. In his private correspondence, Darwin admitted that the *Origin* overestimated the relative importance of natural selection, and by the time he came to write *The Descent of Man* even Darwin's tone was Lamarckian.[26] However, by this time many of those features that we call Lamarckian actually formed part of the Darwinian lexicon,[27] in which a range of different emphases of factors in inheritance were possible. Natural selection and the inheritance of acquired characters had become complementary processes in a Darwinian theory of inheritance. Thus it is a mistake, for example, to label Herbert Spencer as a 'non-Darwinian' or 'Lamarckian' when '[g]iven the indeterminacy over the mechanisms of evolutionary change within Darwin's theory, Spencer's claim [to be defending Darwinism], viewed historically, was quite plausible'.[28] The 1899 rediscovery of Gregor Mendel's pea-fertilization experiments, combined with August Weismann's discovery of an environmentally immune 'germ plasm' – demonstrated by the inter-generational chopping off of mice tails – were indicative of a *fin-de-siècle* shift towards hereditarianism. But even in the Edwardian period the term 'Darwinism' was still broad enough to encompass a predominantly environmentalist explanation of human character.

Equally indeterminant was Darwinism's attitude towards progress. The import of Darwin's theory was that evolution was an open-ended process. Yet in both the *Origin* and the *Descent*, Darwin could not resist occasionally equating evolution, for man at least, with 'progress towards perfection' and the triumph of virtue. According to Bowler, Darwinism was caught between the non-directional logic of Darwin's biological theory and the linear, progressionist accounts of contemporary anthropology and archaeology.[29] In the absence of fossil evidence for biological change, Darwinists seized on the evidence for cultural progress. Anthropological studies by Edward B. Tylor and Lewis H. Morgan, combined with the archaeological findings of Charles Lyell and John Lubbock, encouraged Darwinists, and occasionally Darwin himself, to lend Darwinism a sequential and progressionist view of history. Few accepted T. H. Huxley's image of nature as the inexorable grinding of a directionless machine. Far more copied Spencer's picture of a purposeful universe, in which humans advanced through a pre-determined sequence of stages.[30] Ironically, Spencer's was the more obviously 'scientific' interpretation. The term 'evolution' was derived from the Latin *evolutio* and literally meant the unrolling of a pre-determined plan. This embryological association had been further secured in the mid-nineteenth century by von Baer's research on progressive development through increasing complexity. In contrast, Darwin's neutral understanding of evolution was rooted in non-biological accounts of historical development that sought to avoid presenting history as progress.[31] Thus there was clearly both an isomorphic relationship between scientific and sociological accounts and significant room for manoeuvre in incorporating progress into a broadly defined Darwinian framework. While Darwinism may now be understood as having signalled a shift to a neutral understanding of the term 'evolution', in the late nineteenth century the dominant understanding remained embryological and Darwin's own writings contained encouragement for those socialists who insisted that evolution was progressive.

The same was true for those who clung to spiritualism. Darwin had not, as the apocryphal Harrow schoolboy believed, disproved the existence of God.[32] The 'godless universe' that the *Origin* is now held to herald was only surreptitiously proclaimed in its text. Darwin wrote with nothing like the clarity that would later be assigned to his thought. Despite implying a purely positivistic account of human origins, even the concluding chapter of the *Origin* is concerned with the *ways and means* by which a God might operate, rather than his existence or non-existence.[33] Darwin's immediate objective was to prove only that God ruled by law, not arbitrary intervention, and in doing so his tone was far more reverential and far less obviously atheistic than J. S. Mill's essay on 'Nature', written about the same time.[34] Tellingly, Darwin's circumspection, combined with the efforts of friends such as the American biologist Asa Gray to reconcile evolution and Christianity, and the determination of his devout wife to excise atheism from his *Autobiography*, secured a posthumous assimilation into the Anglican establishment with his burial in Westminster Abbey.[35] Clearly Huxley's 'extinguished theologians' were more imaginary than real, despite his encounter with Bishop 'Soapy Sam' Wilberforce in a set-piece debate in 1860. Both Lucas's debunking of the Huxley–Wilberforce encounter and Osvopat and Moore's demonstration of the extent to which Darwin developed his theory within a framework of natural theology point to the mutual interaction, rather than outright hostility, of late-nineteenth-century religion and science.[36] This was particularly important for socialists, many of whom wished to retain a spiritual element in their thought.

Perhaps the most important indeterminacy that socialists were able to explore within Darwinism was an unresolved ambiguity over the unit of natural selection.[37] The *Origin* presented a particularly brutal picture of *intra*-species rivalry and individual selection. But this opened up the question of how moral qualities such as altruism, courage and moral sensibility might develop. Each clearly had benefits for the group as a whole but was potentially lethal for those individuals who first displayed it. Darwin offered two answers. The first was a sophistic argument of reciprocal altruism – the notion that even altruistic or benevolent behaviour is ultimately selfish and inspired by a calculation of reciprocal benefits.[38] The second was to shift the unit of selection to the group. In such *inter*-group struggles, those groups with 'a high degree . . . of patriotism, fidelity, obedience, courage, and sympathy . . . always ready to aid one another, and to sacrifice themselves for the common good, would be victorious over most other tribes; and this would be natural selection'.[39] This passage from the *Descent* not only 'opened up variable interpretations of the unit of selection', but also implicitly suggested that social qualities were positively beneficial in the struggle for life. Thus chapters III, IV and V of volume I of the *Descent* – which outlined instances of sociability amongst animals and suggested that man's social qualities were the chief factor in his evolution – became particular favourites among socialists. These chapters suggested the clear potential for developing a Darwinian argument against individualism, by suggesting that evolution necessitated man's becoming ever more co-operative and collectivist.

The Descent of Man

Darwinism did not settle questions of heredity and environment, progress and evolution, spiritualism and atheism, or individualism and collectivism. Rather it opened these areas up for further exploration. Darwin raised more questions than he answered. It was in these unresolved areas that a space existed for a genuine dialogue between Darwinism and the left. Even the publication of Darwin's eagerly anticipated anthropological work, *The Descent of Man, and selection in relation to sex* (1871), failed to provide the closure Darwin had hoped for. To his evident dismay, the *Descent* first drew political condemnation from conservative critics, outraged that such a book could be published so soon after the Paris Commune, and then reinvigorated the debate on the implications of Darwinism for understanding man.

Darwin himself had seen enough and retreated from his already limited forays into public life. Sick of the perpetual wrangling, he found solace in that study of earthworms to which he devoted the final years of his life[40] – but not before he had further muddied the waters. Not only had the *Descent* failed to close down the Darwinian debate, it had added two further complicating elements. The first of these, found in the book's subtitle, was the notion of sexual selection. The second was to make an ability to explain distinctions of gender and race a crucial test in the veracity of his theory. After the publication of the *Descent*, questions of race and gender, no less than those of heredity, progress, atheism and the unit of natural selection, came to form part of Darwin's challenge.

Sexual selection, which Darwin first referred to in the *Origin* but developed significantly in the *Descent*, was a waste-bin for collecting up all the odds and sods that did not fit into his theory of natural selection. Some characteristics – such as man's lack of body hair, the horns of a stag and the elaborate plumage of the peacock – could not confer any advantage in the 'struggle for existence'. Special creationists attributed such aesthetic beauty to God's artistic eye, but Darwin, who by natural selection had helped demolish the notion of 'God the architect', was not about to concede the existence of 'God the artist'. Instead he supplemented natural selection with a notion of sexual selection. Each creature was its own artist, as the males and females of each species sculptured the anatomical figures of the opposite sex by their own notions of beauty. Darwin's attempts to prove this – by daubing, docking, and damaging birds in a whole variety of ways to see if clipping their feathers, or dying them another colour, made them more or less successful in attracting mates – were unconvincing and lacking in precision. Nonetheless, the importance of sexual selection was that it represented another retreat from natural selection as the all-encompassing explanation of evolutionary variations and, as with the chapters on sociability, opened up evolutionary alternatives to an individualistic, Malthusian struggle.[41]

Less encouraging were the *Descent*'s strictures on inequality, which betrayed a clear assumption that women and non-Europeans were at a lower evolutionary stage than European men. Darwin demonstrated female intellectual inferiority by comparing lists of the most eminent men and most eminent women in any discipline. His dismissal of the female ability to outperform men in tests of intuition, rapid perception and imitation as 'characteristic of lower races' was doubly revealing. The inferiority of both women and non-Europeans was comparable because both were losers in the process of natural selection. Whereas men had developed 'higher' mental powers through their struggle to maintain themselves and their families, women had been cosseted from the struggle and left concomitantly mentally stunted. Similarly, the ability of Europeans to conquer other peoples suggested that Europeans had honed their superior intellectual faculties in a struggle for life, which rendered them 'fitter' than non-Europeans. Theoretically it may have been possible to overcome such inequalities but other factors would militate against any game of evolutionary catch-up. In the case of women, male sexual selection of the aesthetically pleasing, rather than the most intelligent partners, literally bred its nemesis in the lower intelligence of future generations of women.[42] For non-Europeans, their fate, in what the subtitle of the *Origin* called 'the preservation of the favoured races in the struggle for life', could be seen in the analogy of the New Zealand rat eliminated by the competition of its superior European cousin.[43]

Darwin did not share the harsh racism of some later social Darwinists. And his sexism was probably no greater than that of his immediate contemporaries. But however 'liberal' Darwin's own views may have been, the real significance of his treatment of these issues in the *Descent* was to change the parameters of political debate. Henceforth questions of race and gender were to be debated in biological terms. And to the extent that socialism was forged in an attempt to probe, to explore and to resolve the questions Darwin set, questions of race and gender were correspondingly crucial in the shaping of modern socialism. The indeterminacies of Darwinism meant that there was always more than one answer to any given question. Darwinism did not necessarily favour left or right. What it did demand was that whatever answer one gave must be compatible with biology, and expressed in an organic and evolutionary language.

Darwin's haughty dismissal of J. S. Mill's purely philosophical case for sexual equality – because it ignored modern biology – was a straw in the wind for the reconfiguration of the discursive space in which political debate would henceforth take place. For the few existing socialists this, rather than the details of Darwinism, was what was problematic about the new doctrine. Owenism, for example, had been rooted in an Enlightenment language, which took machinery as its defining metaphor, and had little in common with a biological discourse. Some socialists opted out or, rather, as with Marx and Engels, remained in an alternative discursive space. But later socialists and socialist movements who wished to connect to the politics of their day had to work within the Darwinian framework. For the most part, this made late-nineteenth-century socialism a biological discourse, modelled on the incremental development of an evolving organism – and this, in turn, affected its politics. The first socialist to attempt to meet Darwin's challenge and explore the discursive boundaries of Darwinism was Alfred Russel Wallace.

2

Alfred Russel Wallace

'I hope you have not murdered too completely your own and my child.'
Darwin to Wallace, 27 March 1869

Alfred Russel Wallace is the forgotten man or the 'perennial afterthought in the Darwinian story'.[1] But it was the receipt of his paper, 'On the tendency of varieties to depart indefinitely from the original type', in February 1858, which panicked Darwin into going public. Born in Usk, Monmouthshire in 1823, Wallace was fourteen years Darwin's junior and his formal education had ended when he left Hertford Grammar at the age of thirteen to be apprenticed as a surveyor in his brother's firm.[2] Whereas Darwin studied Paley at Cambridge, Wallace, a restless auto-didact, taught himself from the writings of the socialist Robert Owen and the phrenologist George Combe. Owen's belief in equality and Combe's passion for the structure and workings of the human brain were to remain with Wallace for the rest of his life and profoundly influenced his interpretation of Darwinism.[3]

But it was a third author, Robert Chambers, who was to trigger Wallace's interest in evolution. For the impressionable 21-year-old the *Vestiges* was both an inspiration and an incitement. He was immediately convinced of the case for evolution and was fired with the enthusiasm to discover evidence for how evolution might work. Three years later, in 1848, he got his chance when his friend Henry Walter Bates, a Leicester hosiery manufacturer and amateur naturalist, financed an 'audacious field trip' to find evidence for man's evolution. Four years in the Amazon ended in disaster, however, when the ship carrying Wallace's notebooks and specimens back to England was destroyed by fire.[4] Undeterred, Wallace undertook further fieldwork in Malay, earning his keep by sending specimens back to more illustrious naturalists, including Darwin. Whilst there Wallace contracted malaria and it was when recovering from this latest misfortune that he penned the paper that gave Darwin palpitations.

From his sickbed in Malay, Wallace had independently hit upon the idea of evolution by natural selection, after reading Malthus.[5] In Down House, Darwin wavered over what to do. At one point he began a letter giving up his claim to priority and feared that 'all my originality, whatever it may amount to, will be smashed'.[6] But he need not have worried. Darwin enjoyed powerful patrons in the shape of Joseph Hooker and Charles Lyell. Lyell steadied Darwin's nerves and, with Hooker, arranged for a joint presentation at the Linnean Society on 1 July 1858. An extract from Darwin's unpublished 1844 manuscript on evolution was to be read, along with an 1857 letter that Darwin had sent to Asa Gray. With Darwin's priority thus established the Society then heard Wallace's paper. Wallace himself was neither consulted over the arrangements nor able to attend.

Although Darwin occasionally felt pangs of guilt – amidst all his other ailments – over these proceedings, Wallace, a man of unfailing modesty, never complained about his treatment and soon began publicizing his own discovery as 'Darwinism'.[7] But his fortitude should not be mistaken for deference and Wallace's contribution to the history of Darwinism did not end, as some have assumed, with the reading of his paper at the Linnean Society. Wallace led the way – initially with Darwin's support – in applying evolution by natural selection to the study of man. In particular, his 1864 paper 'The origin of human races and the antiquity of man deduced from the theory of "Natural Selection"' was a seminal moment in the history of Darwinism. It was the first significant attempt to explore the troublesome implications of the *Origin*'s enigmatic promise that '[l]ight will be thrown on the origin of man and his history'.[8]

True to form, Darwin had been reluctant to publish on such a controversial subject but was keen for Wallace to test the waters, even offering the use of his own notes on the subject.[9] But the 1864 paper was not entirely to either man's satisfaction and Wallace gave the matter further consideration in an 1869 review of the tenth edition of Lyell's *Principles*.[10] His conclusion, that man's moral and intellectual faculties were derived not from natural selection but from 'an unseen universe of the Spirit', appalled Darwin who scribbled an emphatic 'NO', underlined three times, on his copy.[11] Thereafter Wallace's Darwinism and Darwin's version were divergent creeds, with Darwin shedding his inhibitions, partly in response to Wallace, and publishing the *Descent of Man*.

Wallace, however, was not to be deterred. His 1864 paper was the beginning of a journey to discover the full implications of evolution by natural selection for mankind. In the process he became first a spiritualist and then a socialist. While Darwin and some historians have assumed this signalled a digression from his science, for Wallace his three 'isms' – spiritualism, socialism and Darwinism – were part of one unified explanation of what his 1870 paper called 'The limits of natural selection as applied to man'.[12] In this paper Wallace confirmed his spiritualism but it was not until he began styling himself a socialist, in 1889, that his interpretation of Darwinism achieved the intellectual coherence he craved. Both spiritualism and socialism were taken up in an exploration of the indeterminacies within Darwinism and both helped Wallace tie up some loose ends. Spiritualism explained the origin of man's moral and intellectual nature. Socialism explained the direction of evolution and the process by which it operated among civilized men. These solutions were more kindly and collectivist than the virulent and individualist set proffered by Darwin's unofficial spokesman, T. H. Huxley. But they were developed within the same Darwinian paradigm and expressed in the same Darwinian language. For Wallace, socialism developed out of Darwinism, not in contradiction to it.

Natural selection and the race debate

It is significant, given the later association of 'social Darwinism' with racism, that the primary purpose of Wallace's 1864 paper – the first significant 'social' application of the theory of evolution by natural selection – was to refute the argument for racial inequality. The contemporary race debate had polarized around a dichotomy between monogenesis and polygenesis. Monogenesis posited a single origin for all of mankind, ascribing racial differences to local variations in the physical climate. This position, which was rooted in the Christian story of creation, was promoted by the Ethnological Society, of which all the leading evolutionists, Darwin and Wallace included, were members. Polygenesis, by contrast, maintained that racial groups had distinct origins and were in fact different species. This newer, ostensibly more scientific, doctrine had been championed by Robert Knox in his influential work *The Races of Man* (1850) and was promoted by his disciple James Hunt, President of the Anthropological Society.[13] Wallace's paper, which was read before the rather unpromising swine of the Anthropological Society, anticipated Darwin's argument in the *Descent of Man*. Both men expressed the hope that a general acceptance of natural selection would allow the monogenesis/polygenesis debate to 'die a silent and unobserved death'.[14]

There was, Wallace admitted, contradictory evidence. Monogenesis could find support in miscegenation, the inter-breeding of different races. This clearly demonstrated that races were not distinct species, but differed only superficially and were probably descended from a single prototype. Polygenists, however, could point to the fixity of racial characteristics under varying environmental influences. For example, even after hundreds of years the descendants of European immigrants to South America had still not developed the look of the natives. For Wallace, the most likely explanation for this contradictory evidence was that a profound rupture had occurred in man's evolutionary development. Different races had evolved from a common stock, and diverged in response to external environmental stimuli – those in hotter climates, for example, becoming darker skinned. But at some point the human body had stopped responding to such stimuli. This was Wallace's point of departure into a new evolutionary theory.

The fixity in man's racial characteristics – which had lasted four to five thousand years – was read as part of a more general exemption of man's bodily form from the physical pressures of natural selection. This exemption, according to Wallace, coincided with the advent of man's 'wonderfully developed brain, the organ of the mind'.[15] From this point on, man met environmental pressures not by changes in physical form but by utilizing his intelligence. Food was gathered by building bows and developing farming techniques, not by growing longer nails or gaining bodily strength. Equally, man survived the glacial epoch by wearing warm coats and building houses rather than growing fur. The brain kept man in harmony with nature, 'not by a change in body, but by an advance of mind'.[16] Natural selection still occurred but it operated on man's moral and intellectual faculties of the mind, not his physical form.

In thus transcending the monogenesis/polygenesis debate, Wallace had developed the theory of evolution by natural selection in three ways. Firstly, he had introduced distinct phases of physical and mental evolution. With the advent of the human brain, physical evolution stopped and mental evolution began. Wallace rooted this rupture in man's sociability. Natural selection worked on the bodies of animals because they were self-dependent and isolated, with little mutual assistance among adults and little by way of a division of labour. Man was different. Even the rudest tribes assisted the sick and practised some division of tasks, which militated against individualism. By thus freeing each other from the daily struggle for life, social co-operation exempted man from the physical pressures of natural selection, allowing the weak and the dwarfish to live and rendering bodily adaptation unnecessary.[17]

Secondly, with natural selection operating on man's mental rather than his physical characteristics, Wallace made the unit of human evolution the group rather than the individual. Morality and intelligence, which lifted man beyond the purely physical demands of animal natural selection, were inherently social attributes. Thus natural selection could occur only through a struggle waged between social groups. Not that this mental natural selection was any less cruel or Malthusian than its physical counterpart. According to Wallace, it entailed the inevitable extinction of all those 'low and mentally undeveloped populations with which Europeans come in contact'.[18] Where it differed, and this was Wallace's third development, was that group selection clearly favoured the evolution of ever more social characteristics. While the Malthusian struggle amongst animals encouraged a rampant individualism, among humans it had precisely the opposite tendency. The more social and sympathetic the human group was internally, then the greater its ability to adapt to its external environment, and the greater its ability to survive in the struggle for life. Thus natural selection implied the triumph of the most collectivist social groups.[19] But these innovations did not satisfy Wallace because they raised two problems, which drove his future speculations.

Spiritualism

The first problem was to explain *how* man's moral and intellectual capacities, upon which the dichotomy between physical and mental evolution depended, had originated. Although he had suggested that such capacities were developed through a process of natural selection, analogous to physical struggle, Wallace had not addressed their source. He had, however, implied that human evolution was qualitatively different from that of animals. This contrasted with Darwin, who in the *Descent* had found man's morality and intelligence foreshadowed in the animal kingdom and saw their human expression merely as an incremental accentuation of animal traits, strictly analogous to man's physical development. Wallace disagreed and was doubtful that man's evolution either blurred so easily into the animal kingdom or could be explained purely in terms of natural selection, or any other type of selection for that matter.

Wallace's interest in phrenology – a science he continued to champion until the end of his long life in 1913 – had taught him to think of the faculties of the brain as discrete psychological units, rather than a gradual accumulation of traits.[20] And not only did man's brain seem to set him apart from other animals, but certain physical characteristics – his upright posture, smooth skin and expressive features – would have been positively harmful in his struggle for life.[21] Natural selection worked by accentuating characteristics of immediate utility in the struggle for existence. Yet man had developed physical and mental capacities way beyond his needs. The hand of prehistoric man was as complex as that of civilized man, even though it had far fewer tasks to perform. Savages enjoyed the same vocal range as civilized man, even though they had never developed a language that required it. Most problematic of all, the human brain, which was essentially the same in savage and civilized men, had a latent capacity far greater than any civilized society had yet realized. Among savages the brain was an organ developed 'quite disproportionate to his actual requirements'. This troubled Wallace deeply. Natural selection would have endowed the savage with an ape-like brain – there was no utility in man's ancestors having had mathematical ability, aesthetic appreciation or moral qualities – yet he actually possessed 'one but very little inferior to that of the average members of our learned societies'.[22]

All this would have been less problematic had Wallace been prepared to countenance evolutionary explanations other than natural selection. Even Darwin had utilized sexual selection as a dumping ground for all the inconvenient facts that did not fit his theory. But Wallace, as he explained in his magnum opus *Darwinism, an exposition of the theory of natural selection with some of its applications*, stuck resolutely to a position of 'pure Darwinism'.[23] The object of this work was to overcome one of the weaknesses of the *Origin*, by finding evidence for natural selection in nature, rather than relying, as Darwin had, on examples of domesticated animals and cultivated plants. In the process, Wallace argued for the 'overwhelming importance of natural selection over all other agencies in the production of new species' and made 'Natural Selection supreme to an extent which even Darwin himself hesitated to claim for it'.[24]

Whereas Darwin increasingly diluted the agency of natural selection in evolution, by suggesting alternative explanations of inheritance, such as acquired characteristics, sexual selection, 'chance' and correlated growth, Wallace was aligned with Weismann in having no time for Lamarckian theories of acquired characteristics and gave little consideration to sexual selection. Wallace wanted to explain everything in terms of natural selection. But man's moral and intellectual faculties did not fit. The sporadic character of man's moral and intellectual faculties, the fact that they were only well developed in a small proportion of the population, even in civilized nations, and the speed with they could develop following civilization were all 'totally inconsistent with any action of the law of natural selection'.[25] This was a problem that Wallace eventually solved whilst sitting around a table, in a darkened room, holding hands and listening intently for signals from the dead.

The American fad of table-turning and spirit-rapping had found a burgeoning British audience in the 1850s and 1860s among former Chartists and ex-Owenites. At precisely the moment when Darwinism might have given a decisive support to their unbelief, atheists were mutating into spiritualists and attempting to summon up the spirit of Feargus O'Connor. Wallace was part of this trend and began attending séances with Mrs Marshall, London's leading spiritualist, in 1865. The attraction of spiritualism for many on the left was that it provided a 'democratic epistemology' at a time when elitist experts were dividing and monopolizing other areas of knowledge.[26] The simplicity and accessibility of the séance was far more congenial to Wallace than the professionalization of science in which Huxley was engaged. And whilst attending one of the séances, which so dismayed Darwin, Wallace hit upon an explanation for the human brain capacity far exceeding the requirements of immediate utility. If not natural selection, man's moral and intellectual faculties must 'have had another origin; and for this origin we can only find an adequate cause in the unseen universe of Spirit'. Not only that, but the brain's latent capacity suggested man was progressing to a yet 'higher' state.

Thus in one fell swoop Providence, teleology and progress were reasserted. Yet in a curious way this was consistent with, and a logical progression from, Wallace's 'ultra-Darwinian' position. It was precisely because natural selection was the only *natural* cause of major evolutionary change that Wallace felt the need to go beyond a purely materialistic explanation of man's evolutionary development. Darwinism 'carried to its extreme logical conclusion', he asserted, 'lends a decided support to, a belief in the spiritual nature of man'. For while it showed 'how man's body may have been so developed from that of a lower animal form under the law of natural selection', it could not explain the origin of man's 'intellectual and moral faculties'.[27]

Socialism

Spiritualism, therefore, accounted for the origin of man's moral and intellectual faculties but it did not explain how their subsequent development was to be guaranteed. This was Wallace's second problem. In his 1864 paper, Wallace had made man's sociability both the cause and consequence of his evolution. It was man's social character, and the resultant division of labour, which enabled his moral and intellectual faculties to develop. Equally, as those faculties developed, so man became more sympathetic and altruistic. Consistent with his belief in natural selection, however, this apparently benign process manifested itself in a fierce Malthusian struggle in which 'low and mentally undeveloped populations' would lose out to 'the wonderful intellect of the European races'.[28] Wallace never doubted the first part of this analysis, that natural selection was the mechanism for the evolution of man's moral and intellectual faculties. What he did come to reject was his own lazy confusion of the 'low' with non-Europeans. This, in turn, led Wallace into a critique of civilized society that culminated in his embrace of socialism as both the guarantor and the endpoint of human evolution.

A confluence of factors set Wallace on this path. Politically, the unsolicited use of his arguments by the *Economist* editor W. R. Greg, in a virulent piece condemning social reform, shocked the old Owenite.[29] So did the feverish race debate, triggered by the concurrence of the American Civil War and the Governor Eyre controversy, which engulfed Britain in the mid and late 1860s.[30] Personally, Wallace's experience of 'savage' people had been positive: 'The more I see of uncivilised people the better I think of human nature', he wrote while living among the Dyaks in Borneo.[31] Spiritually, his view that an 'unseen universe of Spirit' had provided all humans, savage and civilized alike, with a brain of similar latent capacities, was as strong an argument for racial equality as the Christian community of souls. Most important, however, was Wallace's science – in particular, his assumption that the advent of the human brain constituted a rupture in evolution comparable to only two other instances, the divisions between inorganic and organic matter and between the animal and vegetable kingdoms.[32] If this was so, then savage and civilized societies were on an evolutionary continuum but savage and animal societies were not. This separated Wallace from Darwin, and most other

Darwinists, who used evolutionary theory to emphasize how animal society evolved into savage society. Wallace, by contrast, used evolutionary theory to stress the connections between savage and civilized man.

Darwin's position was philosophically more radical but Wallace's was politically more dangerous. By blurring the evolution of animals into savages, Darwin confirmed a set of racial prejudices and encouraged the view that non-Europeans were little better than animals. The slang term 'monkey theory' captured the implication that in Darwinism some races were further advanced than others from their ape-like origins. The subsequent dehumanizing of certain races was not restricted to non-whites – L. P. Curtis has shown how the stereotypical portrayal of the Irish changed from drunken 'Paddy' into dangerous ape man – but non-Europeans were most frequently presented as just above the orang-utan in the evolutionary scale.[33] However philosophically radical, most Victorians – Darwin included – were clearly more comfortable musing on the simian ancestors of man than in considering their own savage roots.[34] Darwin's 'astonishment' on first sighting a party of Fuegians permanently jaundiced his views and he subsequently suffered from the 'English disease' of thinking more highly of animals than of his fellow (non-white) humans. Wallace, by contrast, jettisoned the anthropomorphic assumptions of his early writings and avoided the sentimentality for animals that plagued Darwin's work on sexual selection and facial expressions. Wallace's rigid dichotomy between human and animal evolution may have represented a philosophical retreat but it rammed home the unsettling assumption that white and non-white were on the same side of the evolutionary divide. Not only were other races not to be treated as animals, but savage societies might have something to teach the civilized.

This critical perspective had been lacking in Wallace's 1864 paper. At the time, Wallace had been deeply influenced by Herbert Spencer's 'stage theory' of social development and presented an account of the development of human societies which ranked races and saw competition eliminating the less well adapted. This was to become standard among evolutionary scientists. Darwin marked Wallace's comments heavily and scribbled approvingly: 'natural selection is now acting on the inferior when put into competition with the New Zealanders – high New Zealander[s] say the [Maori] race dying out like their own native rat', a point to which he returned in the *Descent*.[35] Huxley too was convinced that Darwinism justified racial hierarchy and colonial genocide.[36]

But such assumptions were not necessarily Darwinian. Wallace, Darwin and Huxley had all been seduced by an ordinal and progressionist account of group development which owed far more to embryological versions of evolution – such as pre-Darwinian recapitulation theory and the linear models of anthropology and archaeology – than it did to a non-directional, strictly Darwinian account. Yet such accounts segued so neatly into notions of natural selection that only Wallace could see the join. And even he was not prepared to junk all aspects of the model. The linear and progressionistic assumptions were retained. But whereas Darwin and Huxley deployed them as an apologia for existing social and racial relations, Wallace turned them into an instrument of critique.

Instead of unthinkingly placing English society at the top of the evolutionary tree, he argued that the evolutionary process had gone awry. In Wallace's hands evolutionary theory ceased to act as a rationalization of what was and became a promise of what could be. The key here was to hold up so-called savage societies as occasionally more civilized and more advanced than the West. Thus towards the end of his popular travel book *The Malay Archipelago* (1869), Wallace favourably contrasted primitive morality with the 'social barbarism' of Victorian England. If a savage society could attain a higher level of morality, then something must have disturbed England's evolutionary progression. The villain was *laissez-faire* individualism. Human evolution – the development of man's moral and intellectual faculties – depended upon the extent to which man was exempted from an individualistic, physical struggle. Yet Victorian society celebrated individualism, and indeed Greg, and others who were 'totally ignorant of what they are talking about', advocated *laissez-faire* as the means to natural selection. This may have been appropriate for 'infinitely lower level of beasts that perish' but was at odds with the ever more collectivist direction of human evolution.[37]

Thus when Wallace began to style himself a socialist in 1889, it merely confirmed a view he had already reached through his understanding of Darwinism. The direction of evolution was towards an ever more socialistic state, where greater co-operation and division of labour would allow the higher development of man's moral and intellectual faculties. Far from socialism and Darwinism being 'uneasy bedfellows',[38] Wallace made them interdependent. The application of his Darwinism had led Wallace to argue that socialism was the precondition for, and the legitimate outcome of, the evolutionary process.

Only in his later writings did Wallace make this connection explicit. Evolutionary advance depended upon man being ever more social and, as man advanced, so he became more social and less individualistic. Evolution in the civilized world had been corrupted by the spread of *laissez-faire* individualism. The remedy was socialism. What was required was a cleansing of 'the Augean stables of our present social organisation' to rid it of individualism. Once a more socialistic system was in place, 'a system of *truly natural* selection will come spontaneously into action'.[39]

This was not all. In his final work, written in 1913, Wallace took his argument one stage further by making the emancipation of women a crucial precondition for socialism and, therefore, evolution. True natural selection, he argued, required that women be free in their selection of sexual partner. This process had been perverted by a social system that stripped women of all rights and encouraged them to marry for financial gain rather than sexual emotion. Once this was rectified then marriages to the idle, the utterly selfish and the insane would diminish and women's higher social status as sexual selectors would make them the future 're-generators of the entire race'.[40] Thus at the height of the suffragette protests, Wallace connected Darwinism with the case for profound political and social reform. Wallace's socialist exegesis of Darwinism, which had begun with an argument for racial equality, ended in a call for female emancipation.

Legacy

Throughout it all Wallace had retained a 'view of man and society which was still, in essence, *naturalistic*'.[41] His socialism was developed within Darwinism and expressed in a Darwinian language. Perhaps if Wallace had enjoyed a fully worked-out position in the 1860s he might have provided a template for future socialists. As it was, a fair few others had reconciled socialism and Darwinism before Wallace got there in 1889. Moreover, Wallace's dependence upon such esoteric ideas as phrenology reduced his appeal, as did his refusal to countenance Lamarckism. But in other respects Wallace was more typical of the Darwinian left. His belief that man represented a qualitative break from evolution in the rest of the animal kingdom and could, to a certain extent, conquer nature, was to be widely endorsed. So was Wallace's neat side-stepping of the Malthusian problem, by finding the mechanism of natural selection appropriate for animal nature but not for human society and his integration of Godwinian optimism into the process of evolution.

Wallace's greatest legacy, however, was the identification of socialism with evolution. This rendered socialism naturalistic and Darwinian. And it fundamentally changed the essence of socialism. Socialism ceased to be an end and became instead a means. In Owenism and Marxism, socialism came about through a once-and-for-all millennial upheaval. The moment of change was cataclysmic and the state of socialism, once achieved, was static. Any socialism developed within Darwinism was necessarily different. It was a gradual process. The course of evolution was towards socialism and the point of socialism was to allow further incremental evolution.

3

From Radicalism to Socialism

Darwinism was profoundly influential in the late-nineteenth-century movement from radicalism to socialism. The 'central tenet' of radicalism – as espoused by Paine, Thelwall, Hodgskin, Shelley and a host of others – was the 'attribution of evil and misery to a political source'.[1] The accompanying critique of political institutions and the contrast that radicals drew between the harmony of nature and the disharmony of political artifice rested on the deistic assumption of a beneficent natural state.[2] Thus whilst some early-nineteenth-century extremist elements and individuals – such as Hewett Cottrell Watson, William Chilton and Robert Knox – had been militantly atheistic, prior to 1859 mainstream radicalism had remained determinedly providential.[3]

In the longer term such an analysis would prove impossible to sustain in the face of Darwinism. In the same way that the Darwinian revolution ultimately entailed the destruction of the intellectual foundations of Christian providentialism, so it also destroyed the philosophical first principles of radicalism. And both for the same reason: Darwin had delivered a deathblow to the concept of design. In neither case, however, was there an immediate shift. As many scholars have pointed out, the notion of a Darwinian revolution is misleading if the term 'revolution' is intended to signal a short, decisive moment of change.[4] Darwinism was more akin to an acid slowly eating into the intellectual supports of a providential worldview.

What the Darwinian revolution signalled was the beginnings of a hesitant movement on the left, out of the discursive space occupied by radicalism and into a new area framed by Darwinian precepts. This involved an abandonment of radicalism's providential view of nature, and the swapping of mechanistic metaphors, in favour an organic and evolutionary language. The process was neither quick nor straightforward. The movement from radicalism to socialism can be thought of as a complex and convoluted play, involving a number of apparently contradictory acts, and lacking a strictly linear narrative. There were elements of continuity and discontinuity, even within writings of an individual author. As we shall see, socialists such as Aveling who thought they had abandoned radicalism were not completely successful at shaking off their inheritance. Nonetheless, however uncertain and incomplete, a change in the basis of left-wing thinking did occur and Darwinism was a decisive influence.

Henry George

Henry George (1837–97) makes an unlikely starting point for the study of this transition. But then he has long cut an incongruous figure as the 'godfather' of British socialism. An 'apostle of frontier individualism and free trade', steeped in the Anglo-American radical tradition that dominated the left on both sides of the Atlantic, George was no more a socialist than he was a Darwinist.[5] He had little interest in science – beyond being imbued with the general evolutionary ethos of the age – and his major work, *Progress and Poverty* (1879), was the final flowering of radicalism; one final restatement of the case against classical economics and one last dismissal of the *bête noire*, Malthus. But probe a little deeper and one can see that *Progress and Poverty* represented a subtle shift in the left critique: the abandonment of deistic and providential assumptions about nature and the first shoots of a new argument, prompted by the challenge of Darwinism. Not that George consciously had any such objective in mind. He sought only to dispute with the long dead classical economists, Ricardo, Malthus and Mill. These, of course, were the enemies of early-nineteenth-century radicalism and both the content and structure of George's argument – proceeding from a critique of the wages fund, through to the population principle and culminating in a critique of property rights – was, accordingly, little more than reheated Hodgskin.[6] The only real innovation was George's 'True Remedy' to the problem of poverty amidst progress – a single tax on ground rent to replace all other taxes – and this never really caught on.[7]

But below the superficial similarities there was a significant difference between George and his radical predecessors. When it came to dealing with Malthus, George had to confront not only the 'Parson' but also his curate, Darwin. This left a simple choice: either tackle Darwinism head on and dispute the veracity of the struggle for existence among animals as well as among men; or sidestep the issue and take man out of nature. George chose the latter course and in doing so recast the discourse of radicalism in a way that dispensed with the need for a providential view of nature and offered a point of departure into a new socialist argument.

The key lay in dichotomizing the human and animal worlds. There was, George maintained, a qualitative difference – 'not merely of degree but of kind' – between man and all the other animals.[8] The ability to progress had put 'an irreconcilable difference' between even the lowest savages and the highest animals.[9] The dawn of human history was the point at which man crossed the Rubicon of progress. This inaugurated 'a progression away from and above the beast'.[10] Man may have been 'only a more highly developed animal', but '[b]y whatever bridge he may have crossed the wide chasm that now separates him from the brutes, there remains of it no vestiges'.[11] Whereas the animal kingdom, George argued, engaged in an endless cycle of satisfying its fixed desires for food, shelter and reproduction, at some long-distant point man had begun to use the fulfilment of these desires as a base from which to strive for improvement. By thus placing man in his own separate sphere of history and culture, George ceded nature – the animal and vegetable kingdoms – to the internecine warfare of the Malthusian struggle. This was radicalism but with a difference. Evil was still the product of political institutions but in the context of the imperfections of human culture, rather than as a perversion of nature. Man's sufferings were still man-made and by man therefore remediable. This, however, was no longer argued on the basis that nature was benevolent, but because man was the master of his own space: culture.

Of course, a strict Darwinian could retort that whatever progress man had made away from a simple bestial existence of food, shelter and reproduction, he was still a slave to his nature in terms of inheritance. Here George fell back on a position that combined a traditional radical defence of the Lockean *tabula rasa* with an added sociological twist of cultural inheritance. Rejecting crude but popular racial theories about progress, like the evolution of the species, being guided by fixed laws carried by inheritance from one generation to another, George pointed out that the most advanced civilizations of previous ages had tended to putrefy and degenerate.[12] Yet even during the decline of ancient Greece and Rome, their populations had consistently produced babies that were as physically fresh and biologically healthy as their ancestors. What had putrefied, and held subsequent generations in check, was not 'natural' hereditary material, but the culture of these states. It was traditions, beliefs, customs, laws and habits – 'the matrix in which mind unfolds and from which it takes its stamp' – that had atrophied and turned progress into decline.[13] In turn, the path of progress had then been taken up by societies that a few generations previously had been regarded as barbarians but whose supposedly inferior biological inheritance was set to nought by a healthy culture. Similarly, the superiority of nineteenth-century men over their ancestors was a product of social and cultural, not biological, inheritance. A set of European babies, left to fend for themselves on a desert island, George declared, would be back to stage one.[14] The differences that existed between men over generations arose 'because we stand on a pyramid, not that we are taller. What the centuries have done for us is not to increase our stature, but to build up a structure on which we may plant our feet.'[15]

This answer to the problems posed by Darwin was to prove immensely influential in the formation of a socialist left. George's two manoeuvres – dichotomizing humans and animals, and restricting the action of evolution to the mind – had been anticipated by Wallace. But it took *Progress and Poverty*, the best-selling economics book of the nineteenth century, and George's promotional lecture tours of the 1880s, for them to become leading strings for the first faltering steps of the new-born socialist movement. Wallace described *Progress and Poverty* as 'undoubtedly the most remarkable work of the present century' and recommended it to an unenthusiastic Darwin.[16] Not that Wallace wholly endorsed either George's philosophy or his programme. Philosophically, George's account of the separation of humans from animals was rooted in a material progress that was closer to Marx and Engels' account than to Wallace's idealistic infusion of an 'unseen Spirit'. Equally, although *Progress and Poverty* focused Wallace's mind on the 'land problem' he soon dismissed George's single tax, preferring the more socialistic solution of nationalization.[17]

The reason Wallace held *Progress and Poverty* in such high regard was that it was one of two books – the other was Edward Bellamy's *Looking Backwards* – which he credited with converting him to socialism. Wallace recognized that in the process of responding to Darwinism, George had inadvertently laid the basis for a new type of left. It was still recognizably radicalism – because the key to human improvement still inhered in changing man's political institutions – but the philosophy underlying this no longer assumed that nature was beneficent. It accepted the operation of natural selection in the animal and plant kingdoms while making man's cultural space distinct from 'lower' nature. This was radicalism secularized. Metaphysical and providential assumptions retained an appeal on the left – George's complaint that it was Malthus who had made Darwinism atheistic found many echoes.[18] But an important turn towards a non-providential, and ultimately socialist, argument had been taken. That this movement was only tentative and contradictory, however, was confirmed in the writings of Edward Aveling.

Edward Aveling

At first glance Aveling (1851–98), a self-titled 'Lecturer in Darwinism', and a Marxist to boot, would seem a far more obvious candidate for effecting a fundamental transformation in the left's philosophy. He certainly possessed the necessary intellectual gifts. Born in 1851 of Irish parents, Aveling was one of the few truly outstanding scientific minds on the late-nineteenth-century left. After gaining his doctorate in 1876 Aveling was honoured by his professional colleagues as a fellow of the Linnean Society, a fellow at University College London and a lecturer in comparative anatomy at the London Hospital.[19] But he never won a corresponding level of respect or influence in his political career. From his first involvement in the secular movement of the late 1870s, through to his service on the founding committee of the Independent Labour Party,[20] Aveling consistently failed to make the kind of impact one might have expected.

Partly this was due to the philandering and money filching that blighted his personal life. So serious were these that Aveling is today chiefly remembered for his repeated betrayals of the revered Eleanor Marx[21] and Bernard Shaw's caustic comment that 'If it came to giving one's life for a cause one could rely on Aveling, even if he carried all our purses with him to the scaffold.'[22] But there was also a deeper reason for his failure. The movement from radicalism to socialism that he hoped to effect was only partially achieved in his own writings. For while Aveling thought his refusal to dichotomize humans and animals made him more truly Darwinian than Wallace, it left him with the problem of explaining how socialism was possible in a Malthusian nature. This was a problem that Aveling solved by reviving the providential assumption of a benevolent nature that George had cast aside.

Unlike George, Aveling saw himself as both a socialist and one of the 'vast and ever-increasing army of scientific children' destined to carry on Darwin's work.[23] He saw no contradiction in the two positions. Indeed, as much as his pupil Annie Besant, Aveling identified a socialist imperative in Darwin's work. It was Darwinism, he argued, that had led him to identify the 'two great curses of modern civilisation', Christianity and capital.[24] As an evolutionist, Aveling explained, he found religion 'a bane and not a blessing'. 'Equally, as an evolutionist, I have come to the conclusion that the present system of production – the capitalistic system of production – is a bane and not a blessing to the world at large.'[25] Darwinism, he said, led him to socialism because it focused attention on the importance of the environment in shaping the development of species. Whereas the positivist limited himself to moralizing the individual, and the radical to recommending a change of government, Darwinian science taught that both were useless unless the environment was also changed.[26] And it was this that made Aveling a socialist.

On Aveling's own account, Darwinism had highlighted the inadequacies of radicalism. As with Besant, who thought the radical 'a half-fledged Socialist', so Aveling liked to think of socialism as an advance on his radical ancestor.[27] He was, he wrote in 1884, 'a Positivist and something more', and 'a Radical and something more'.[28] The 'something more' was an evolutionist and this demanded that the left transcend radicalism. Just as liberalism was an advance on whiggism and radicalism an advance on liberalism, so socialism was the next stage on from radicalism but with one difference. The gap 'between the position of Radicalism and that of Socialism is much greater than between either of the other classes'.[29] In fact, although Aveling's achievement in integrating Darwinism on the left should not be underestimated, he was prone to overstate the decisiveness of his own movement from radicalism to socialism.

This followed from his tendency to see only certainties in the impact of Darwinism and a corresponding failure to recognize the structural indeterminacies in Darwin's legacy. A good example of this was Aveling's crude insistence on an identity between Darwinism and atheism. He was even more sure than J. W. Draper that all science was 'inevitably and essentially irreligious' and saw himself engaged in an 'irreconcilable' mortal combat with religion – an 'internecine warfare' from which science would emerge triumphant.[30] But the majority of Darwinists and the bulk of secularists did not share Aveling's Manichaean mania. Darwin himself resisted Aveling's attempts to drag him into religious controversy, demurely declining Aveling's offer to dedicate an atheistic pamphlet to him,[31] and irritating Aveling by his inclusion in the *Origin* of the odd 'unfortunate phrase' about a creator.[32]

The secularists evinced a similar reluctance. Although Aveling took Darwinism to the secularist movement with a proselytizing zeal that has assured him of a place in the history of freethought, he met with only a muted response. The articles he contributed to the *National Reformer* in the early 1880s, along with his polemical pamphlets and brief editorship of *Progress* during the imprisonment of G. W. Foote, won some converts. But not only were most of his audience wedded to the pre-Darwinian philosophical arguments for the material nature of mind that had sustained them for thirty years; they were also suspicious that Darwinism was being erected into a new religion, with scientists as the new priesthood.[33] Aveling's insensitivity to such charges can be seen in the title of his pamphlet *The Gospel of Evolution* (1881). Moreover, his claim that 'the new evangel is founded wholly on a natural and scientific basis', and assertion that only those with scientific training were fit to speak on complex issues, simply confirmed secularist suspicions.[34]

Equally crude was Aveling's insistence that Darwin had finally settled all questions of continuity between the mental phenomena of animals and humans. He took particular exception to George Romanes' *Mental Evolution in Animals* (1882), which, out of deference to Wallace, excluded man from consideration. For Aveling nothing could be more absurd than Wallace's 'old fancies' about special creations.[35] Darwin had demonstrated 'the unity of all phenomena'.[36] As the title of another pamphlet *Monkeys, Apes, Men* (1884) indicated, Aveling thought there nothing special about men, biologically or historically – even the human mind was merely 'a function of the nervous system'.[37] All things were part of one huge continuum, from inorganic to organic and from plants to animals, including humans. Man had not only evolved from the animals, he remained part of the animal world. Far from being unique, every one of man's mental capacities – intellectual and moral – was anticipated in the natural world.[38]

The benign danger of this argument, as we have already seen in the contrast of Wallace and Darwin, was that it encouraged a sentimental and an anthropomorphic streak. Its malign correlation was the confirmation of the prevailing racial prejudice that non-Europeans were little better than animals. Aveling fell victim to both. In direct contradiction to Marx and Engels, Aveling's *The Origin of Man* (1884) cited examples of tools, fire, dress, houses and property among the 'lower animals'.[39] Progress, which George had found to be uniquely human, Aveling found practised by those birds that adjusted their flight to take account of telegraph wires. Altruism, which Wallace thought uniquely human, was but one instance of the widespread communism that Aveling found in the 'social union between plants and animals'.[40] Indeed, Aveling seemed to hold some animals in a higher regard than he did human savages. There was, he declared, a greater interval between the highest and lowest men than between the lowest man and the highest ape.[41] And 'the virtues of mutual love' were 'shown far more powerfully' by some 'non-human animals' than by black Australians, whom Aveling accused of murdering their children.[42] His criticism that the idea of man's infinite superiority over other creatures arose from 'considering only the European peoples and the contrast between them and the anthropoid apes' was not only unfair to Wallace – who had, unlike Aveling, lived among the so-called savages – but confirmed Aveling's assumption of European superiority.[43]

Not only was the taste of Aveling's Darwinism unpalatable on this point but his assertion that only his was a true and consistent Darwinism was deeply misleading. Darwinism, as we have seen, was not a set meal. Aveling, as much as Wallace, took full advantage of the *à la carte* menu to indulge the peculiarities of his appetite. Just as he overemphasized the decisiveness of the blow Darwin had laid against religion, so he underplayed Darwin's case that an evolutionary differentiation in the mental capacities of the sexes had rendered women inferior to men. The differences in the mental powers of the sexes were brushed aside as a 'discrepancy' to be resolved by opening up educational opportunities, which, 'in the course of time', would ensure that women's brains would equal male capacity.[44] The speed and adaptability this implied was markedly Lamarckian, as was Aveling's whole emphasis on the importance of environment to evolutionary explanations.

It is noticeable that Aveling remained silent on the question of a harsh, unforgiving natural selection. The silence was essential to his political stance. Although Aveling refused to dichotomize man and animal, after placing man in nature he could save socialism from the blight of natural selection only by rendering nature itself benevolent. In doing so, Aveling revived the romantic deification of nature that Wallace and George had eschewed. In *The Gospel of Evolution* (1884) and his pamphlet *God dies: Nature remains* (1881), Aveling's poetical eulogies to nature owed more to his beloved Shelley than they did to Darwin.[45] This highlights Aveling's inability to achieve a clean break with radicalism.

Ultimately, and ironically, even Aveling's atheism was justified in terms of rescuing nature from the aspersions that a God-ruled universe would cast on it. The unworldly pretensions of Christianity were criticized for casting a shadow over man's relation to nature (always significantly with a capital N) and Aveling recommended man turn back to the study of nature – where the 'gospel of evolution' was revealing a new loveliness.[46] Thus for all his dogmatic Darwinism, and his contribution in bringing a new level of scientific understanding to the left, Aveling demonstrated both the contribution of Darwinism to the movement from radicalism to socialism and the hesitant uncertainty that characterized that process.

Kropotkin

Similar problems bedevilled the theory of mutual aid developed by the Russian aristocrat turned anarchist Prince Peter Kropotkin (1842–1921). Kropotkin was almost as incongruous an influence on the development of British socialism as Henry George. But, contrary to Singer's view, neither the exotic origins of this émigré outsider nor his unfashionable anarchism prevented his major work *Mutual Aid: A factor in evolution* (1902) from influencing the mainstream left.[47] Just as importantly, Kropotkin's personal odyssey, from soldier to scientist to socialist and anarchist, provides a good illustration of how science often preceded socialism in the late nineteenth century.[48]

Like Aveling, Kropotkin achieved scientific eminence before he turned to politics. In 1871 he was offered the chance to become secretary of the Russian Imperial Geographic Society but declined in order to dedicate his life to the workers' movement. Nothing in his background suggested such an outlandish course. Aged 15, Kropotkin had followed his father into the military and served for ten years before joining the Tsar's Central Statistical Committee in 1867. The previous five years had been spent in the Siberian service, undertaking military and commercial expeditions, and travelling some 50,000 miles back and forth across the bleak, desolate tundra of eastern Siberia and northern Manchuria.[49]

This period was Kropotkin's rite of passage. He read avidly, including Darwin's *Origin*, and made studies of the geography and species of the area. It was Kropotkin's *Beagle* journey but with one significant difference. Whereas Darwin found an intra-species struggle for existence in the thick, luscious environment of South America, Kropotkin found the creatures of the tundra supporting each other, as they struggled to maintain life and preserve their species 'against an inclement Nature'. What he failed to find, he later wrote, was 'that bitter struggle for the means of existence, *among animals belonging to the same species*, which was considered by most Darwinists (though not always Darwin himself) as the dominant struggle for life, and the main factor in evolution'.[50]

This observation, with which *Mutual Aid* opened, was to form an essential building block in his anarchism. But it was made – as can be seen by the letters he wrote to his brother at the time – long before he embraced the anarchist cause. The notion of mutual aid, as Daniel Todes has shown, was almost a commonplace in Russia, where physio-geographical realities ensured that Darwin's Malthusianism had little resonance.[51] It was only when he was brought into contact with a quite different tradition – English individualism – that Kropotkin felt compelled to elaborate what he had regarded as simple common sense. Kropotkin came to England in 1877, following his daring escape from the Peter and Paul Fortress in St Petersburg, where he had been imprisoned for writing a subversive pamphlet, and spent most of the next forty years living in and around London.

Failing health and the lack of an indigenous anarchist movement led Kropotkin back to his scientific studies.[52] The spark that fired the beautiful unification of his science and his politics in *Mutual Aid* was an 'atrocious article' by 'Darwin's bulldog', T. H. Huxley.[53] Two elements of Huxley's argument raised Kropotkin's ire. The first was Huxley's rendering nature as 'red in tooth and claw' and his understanding of the Darwinian notion of struggle as exclusively individualistic and competitive. This grated with the Russian tradition in which Kropotkin had been educated. Huxley's second, and even greater offence, which he compounded in his subsequent Romanes Lecture, was to dichotomize evolution and ethics as distinct entities. By casting nature as a purely competitive and individualistic theatre, Huxley had assumed that natural, evolutionary behaviour was non-ethical and that ethics was exclusively the province of civilized man in culture.[54] This was to prove a popular argument on the left but it was not one that Kropotkin could accept.

In order to maintain the unity of natural and social worlds, Kropotkin needed both to identify the natural evolutionary roots of ethical behaviour in lower species and to provide an evolutionary mechanism through which such traits could develop over time. The first required very little innovation. Russian naturalists, led by K. F. Kessler, ichthyologist and rector of St Petersburg University, had long argued that Darwin's notion of struggle should not be interpreted in a narrow individualistic sense of animal pitched against animal, or man competing with man. Amidst the huge landmass and sparse populations of Russia, it made more sense to understand struggle in a wider 'metaphorical' sense of a struggle against external conditions, in which species found co-operation to be advantageous to survival. This was the essence of that intra-species co-operation, for the survival of the group, that Kropotkin called mutual aid.[55] Nature was not exclusively the brutal, competitive, individualistic struggle that Huxley claimed. It was also the case that ethical behaviour, in the sense of co-operation for the greater good – the survival of the species – was a natural, instinctual, behavioural trait.

But even though Kropotkin had identified the natural root of ethical behaviour, this still left the problem of how to explain the evolutionary development of such traits. Without such a mechanism, all Kropotkin would have achieved was to highlight static curiosities, such as the co-operation of ants and bees, repeated *ad infinitum* but never developed or augmented over the generations. For an evolutionist this would not do. And for an anarchist it was inadequate. If Kropotkin's conclusions were to carry any political force – and as a Darwinian he saw no distinction between natural and social phenomena – then he needed to show that the natural evolutionary tendency was towards an ever more social and co-operative condition.

This he did by tying together two loose ends in Darwin's writings: the suggestion in the *Descent of Man* that sociability was a factor that had favoured human evolution and the Lamarckian flavour of Darwin's revisions to the sixth edition of the *Origin*. The first showed that the development of co-operation was evolutionarily advantageous; the second provided a mechanism through which mutual aid could be transmitted and developed. Whereas Huxley had confined ethics to civilized society, by building on Darwin's suggestion of a Lamarckian mechanism of development Kropotkin was able to re-unite the natural and social worlds and suggest a pattern to evolution. The chapter order in *Mutual Aid* pointed to a sequential progression in which co-operation amongst animals led on to that among savages, and savage co-operation led on to the co-operation found in medieval societies.[56] In Lamarckian terms, the habit of co-operation encouraged further co-operation. Codes of behaviour, language and a sense of common interest, evolved through the act of co-operation itself.

Kropotkin avoided the pitfall of simplified progressionism by recognizing the possibility of regress as well as progress. Lamarckism was a doctrine of 'use it or lose it'. If the innate tendency towards mutual aid was not fostered and developed then previous advances could be lost through the inheritance of acquired characteristics. Men, that is, could become less social and less co-operative, as well as more. Indeed, this was precisely what Kropotkin claimed had occurred; modern societies showed less evidence of mutual aid than did medieval societies. This was the point at which Kropotkin's evolutionism met his anarchism. The natural tendency to mutual aid had been disrupted by the intervention of the modern state, which usurped many of the functions that should have been performed by society and sent evolution into regress. Thus, for Kropotkin, the theory of mutual aid proved the superfluity of the state.[57]

Few on the British left accepted this anarchist conclusion but many were impressed by Kropotkin's demonstration of the tendency of co-operation to augment with evolution. *Mutual Aid* appealed beyond the anarchist left because it was neat, superficially scientific and ostensibly Darwinian. The syllogism was simple: animals are social: humans are animals: humans will become more social because sociability is an evolutionary advantage. Moreover, the articles that eventually made up *Mutual Aid* gained a wide audience from their publication in the popular journal *The Nineteenth Century*.[58]

There were, however, three weaknesses in Kropotkin's argument. The first – only likely to bother other naturalists – was that despite its subtitle Kropotkin's book had very little to say about mutual aid *as a factor* in the evolution of the physical characteristics of species. *Mutual Aid* demonstrated the advantages of co-operation, and the necessary relationship between co-operation and ethics, but said little about how this contributed to speciation. More problematic was the fact that the theory of mutual aid rested so heavily upon the Lamarckian theory of inheritance. Kropotkin may have been correct to argue that Darwin was more Lamarckian than Huxley cared to admit. However, point scoring was not enough to counter the rise of Weismann's harder-edged theories of inheritance. From the start Kropotkin was forced to fight a rearguard action to protect the Lamarckian principles on which his theory depended.[59]

The third problem was more paradoxical. The most obvious strength of the theory of mutual aid was also its most profound weakness. Kropotkin had answered Huxley's law of competition with his own bio-social law of co-operation. This made *Mutual Aid* impressive but it also carried the danger of any social Darwinist tract: reducing human beings to upright apes determined by their 'nature'. In Kropotkin, concepts of animal and human nature loomed large but the power of 7,000 years of human culture was more or less absent, except in the purely negative sense of the state disrupting the progress of mutual aid. The resultant political prescription – stripping away the 'artificial' accretions that perverted 'natural' development – was all too reminiscent of early-nineteenth-century radicalism.

Conclusion

The cases of Aveling and Kropotkin point to the tentative and often contradictory nature of the movement from radicalism to socialism. But as the example of George shows, there was a certain logic, if not an inevitability, in the shift. It is important to re-emphasize that this was rarely a conscious movement. Radicals did not decide en masse or overnight to become socialists. Nor did socialists exist pre-formed before 1859, ready to emerge to consider the problems thrown up by Darwinism. Two more complex processes were at work. The first was that the ideology of the left was being constantly reassessed and, in a period steeped in Darwinism, exceptional individuals such as Wallace, George, Aveling and Kropotkin were finding the old radical answers unsatisfactory. The second, more typical process was that of generational change. There was no mass loss of faith in radicalism, any more than there was a conscious turning away from the Church. But just as religious observation declined with each new age cohort, so radicalism increasingly rang hollow for each new generation of the left. It was not a case of rejecting radicalism. A Darwinian upbringing meant that one could never really seriously entertain it. Socialism was a different proposition. It was by its very nature – and the timing of its birth in modern form – far more suited to the 'after Darwin' world than radicalism ever could be. It was more biological in its assumptions, more evolutionary in its conception of change, and more organic in its language. However, not all of these lessons were learnt directly from Darwin. An almost equally important influence was Herbert Spencer.

4

'Social evolution is exasperatingly slow, isn't it sweetheart?'

In 1908 the American writer Jack London published his most overtly political novel, *The Iron Heel*. Written after an exhausting year lecturing as President of the Intercollegiate Socialist Society, the book charts the growing tensions, culminating in violent confrontation, between an oligarchy consisting of the government, big business and the military – the 'iron heel' of the title – and the socialistic evolution of society. With opening scenes set in the near future, the tale is told through the eyes of Avis Cunningham, a respectable daughter of the bourgeoisie, who is enraptured by the sublime physicality of the broad-shouldered and firm-muscled proletarian Ernest Everhard.

In order to circumvent the normal literary constraints of the political novel, London adopts two devices. Firstly, the book is presented as a recovered manuscript. This enables London to frame and annotate Avis's story in a second authorial voice, that of Anthony Meredith, a resident of a socialist society seven centuries in the future. Secondly, Everhard's dinner party diatribes, which dominate the early chapters of the book, provide London with the opportunity to deliver an unbroken political analysis. These, however, are among the weakest sections of the novel, which critics, more comfortable with London's less obviously political nature stories, panned. But whatever its literary failings, *The Iron Heel* was a slow-burning sensation in socialist circles, winning praise from as unlikely a trio as Trotsky, Eugene V. Debs and Aneurin Bevan. Taken with London's own immersion in American socialist circles, this suggests that the book might tell us something important about turn-of-the-century socialism.[1]

The formal politics that Everhard and Meredith espouse, however, are less interesting than the imagery and language that London deploys to give Avis's story meaning. Three images, in particular, dominate *The Iron Heel*. The first is of Everhard as 'a superman, a blond beast such as Nietzsche described'.[2] Everything about Everhard, from his name onwards, is designed to suggest qualities of physical and intellectual strength. This Nietzschean imagery is magnified through the adoring eyes of Avis. Her account goes beyond a conventional romance to almost deify

Everhard, who is endowed with Christ-like qualities and whose end is referred to as a 'crucifixion'.[3] Everhard is not simply a hero, he is a superman. This contrasts sharply with the second central image, that of 'the people of the abyss'.[4] The slum-dwellers of Chicago, who rise against 'the iron heel' in the book's concluding chapters, are portrayed as an atavistic and degenerate mob. They are by turns an 'abysmal beast' or 'the monster', who, as Avis is warned, are 'not our comrades'. London borrows heavily from Gustave Le Bon's theory of the crowd to suggest that the people of Chicago are animals, forming a helpless herd and driven by animal instinct and atavistic desire.[5] The contrast between the Nietzschean revolutionaries and the degenerate mob could not have been greater. Both images, however, are only sub-texts in the metaphor that underpins *The Iron Heel*: the battle between 'the machine' of the oligarchy and the evolution of the natural social organism.

As the battle on the streets of Chicago reaches its climax, the true nature of the 'machine civilization' that the oligarchy has built becomes apparent. The allegories flow thick and fast. As Avis surveys the destruction of the city, her comment that 'the stone facing of the building was torn away, exposing the iron construction beneath' carries a deeper meaning. The armed vehicles of the police are referred to as 'the machine' and the mob are mown down by *machine* guns.[6] Against this malign mechanism London pitches a benign, organic evolution, understood only by those who have studied biology and sociology. The weakness of the oligarchy, as Everhard sees it, is that 'they are merely businessmen . . . not biologists nor sociologists' and thus unable to comprehend the 'unseen and fearful revolution taking place in the fibre and structure of society'.[7]

The transition was to socialism, for while one could not predict or rush evolution – 'Social evolution is exasperatingly slow, isn't it sweetheart?' Everhard remarks to Avis – one could grasp its general tendency. And, as in Kropotkin's *Mutual Aid*, the 'tide of evolution' had favoured combination over competition since that moment 'a thousand centuries' ago when primitive man had triumphed over competitive beasts through combination.[8] Even the great industrial and financial trusts that had arisen in competition had combined to destroy it, and socialism, 'a greater combination than the trusts', was next 'in line with evolution'.[9] The strength of the socialists lay in their scientific understanding of the social organism and this sustained them in their fight against the machinery of the 'iron heel'.

London's inspiration was Darwinian, in the broadest sense of the term. Like Kropotkin, London had first given serious thought to Darwinism amidst the icy plains of a desolate landscape. But the lessons London learnt, across the Bering Straits in the gold-rush mania of the 1890s Klondike, were slightly different from those of the Russian aristocrat. Where Kropotkin saw only co-operation, London reflected on the severity both of the struggle with the elements and that between man and man.

During this period, London drank thirstily at the fountain of evolutionary theory, boring his associates by quoting whole passages from the *Origin* and thrilling at Haeckel's *The Riddle of the Universe* (1899).[10] Above all others he was impressed by the writings of Herbert Spencer. And above all else, the central metaphor of *The Iron Heel*, of the machine hopelessly grinding against the course of organic evolution, is Spencerian. Anyone who doubts the debt need only consult *Martin Eden* (1909), the semi-autobiographical novel London wrote about the same time as *The Iron Heel*. Set in San Francisco at the turn of the century, the eponymous hero, a young seaman struggling to attain intellectual and social recognition, is so obviously London that when he describes the electrifying pleasure of reading Spencer we can be sure it is the author's moment of revelation.

> Martin Eden had been mastered by curiosity all his days. He wanted to know, and it was his desire that had sent him venturing over the world. But he was now learning from Spencer what he never had known, and that he never could have known had he continued his sailing and wandering forever. He had merely skimmed the surface of things, observing detached phenomena, accumulating fragments of facts, making superficial generalisations – and all and everything quite unrelated in a capricious and disorderly world of whim and chance.
>
> . . .
>
> And here was the man Spencer, organising all knowledge for him, reducing everything to unity, elaborating ultimate realities, and presenting to his startled gaze a universe so concrete of realisation that it was like the model of a ship such as sailors make and put in glass bottles. There was no caprice, no chance. All was law. It was in obedience to the same laws that the bird flew, and it was in obedience to the same law that fermenting slime had writhed and squirmed and put out legs and wings to become a bird.[11]

In the spirit of Chambers, more than Darwin, Spencer provided his auto-didactic readers with a cosmological view uniting all aspects of the universe in the same evolutionary framework.

Herbert Spencer

London was far from unique among socialists in finding an inspiration in Spencer's writings. The man Beatrice Webb described as 'England's greatest philosopher' was the most popular philosophical writer of his generation and was avidly read by left and right alike.[12] Today his long, dry and repetitive books lie unread, gathering dust on library shelves. But in the 1890s, when a million copies of his various works were in circulation worldwide, Spencer was essential reading for educated men and women.[13] More than any other individual, Darwin and Huxley included, it was Spencer who made evolution the defining idea and the master discourse of the late nineteenth century. It was Spencer who coined the phrase 'the survival of the fittest' – which both encapsulated and helped shape the *zeitgeist*.

Even before Darwin started to write the *Origin*, Spencer had begun to teach how to think about society in organic and evolutionary terms. It was precisely this mode of thinking that Spencer gave to socialism. As the quote from *Martin Eden* suggests, Spencer provided a cosmology; a unifying principle of 'cosmic evolution' that was applicable to all areas of social and natural science: from the development of species to the evolution of the solar system and the maturation of the embryo to the course of social development. This faith that the totality of the human experience could be explained by the same set of evolutionary laws broached no distinction between the natural and social sciences. Spencer's life work was to construct what he called his 'synthetic philosophy', uniting studies of biology, ethics, philosophy, psychology and sociology in one grand evolutionary framework.[14] And he inspired socialists with the desire to do the same.

Wallace was so impressed with Spencer that he called his eldest son Herbert Spencer Wallace. This action, which drew a rare piece of sarcasm from Darwin's pen, might seem perverse given Wallace's intellectual trajectory. After all, for the English Marxist Henry Hyndman, Spencer was 'the favourite philosopher of the successful railway man and stock-jobber', and even a superficial glance at his writings reveals a commitment to *laissez-faire* at odds with Wallace's growing socialist sympathies. From his first published writings on 'The proper sphere of government'(1842) through to his defining work, *Man versus the State* (1884), Spencer had virulently declaimed against all social legislation, even such apparently commonsense measures as sanitary provision and universal education.[15] He was an inveterate opponent of the state and saw socialism as both morally anathema and, worse still, antagonistic to evolution. In Spencer's view, socialists ignored the fact that the condition of society was determined by evolution and the level of human nature, which could only be slowly transformed by natural laws, refractory to legislative fiat.[16]

Below this superficial antagonism, however, Wallace, and other socialists, came to see a deeper intellectual compatibility. Partly this was a product of their shared heritage. Not only did Spencer's philosophy grow out of a radical non-conformity, mixed with phrenology, Chambers' teleology and a Lamarckian emphasis on co-operation. But there were even exceptional issues, such as land nationalization and anti-Malthusianism, where Spencer anticipated the socialist position.[17] Most importantly of all, Spencer's system of thought, like that of Darwin, was shot through with the kind of indeterminacies that left it open to revision and reinterpretation.

Two elements, in particular, of Spencer's synthetic philosophy became entangled in the development of socialist thought. The first, as we have seen in *The Iron Heel*, was the notion of society as a social organism. The second was Spencer's model of historical development, which unintentionally provided a powerful argument for a natural progression to socialism. Whether any of this was strictly Darwinian is about as relevant as asking how many angels can dance on the head of a pin. There is quite a trade in historians disputing the finer of points of what makes a social Darwinist and some of these have argued that it is a mistake to include Spencer in this category, preferring 'social Lamarckian' for Spencer, or even 'social Spencerian' for his followers. But, as Mike Hawkins has shown, the case for not including Spencer is overstated.[18]

It is true that Darwin disliked Spencer, whom he found haughty and unreadable, and that Spencer was prone to tetchily assert his own priority in applying evolutionary theory to human character and society.[19] Nor were their grounds for disagreement purely personal. Spencer denied the legitimacy of the analogy between artificial and natural selection upon which Darwin's *Origin* was premised and argued that natural selection among the higher organisms was increasingly displaced by a Lamarckian inheritance of acquired characteristics.[20] Moreover, Spencer's assertion of the necessity of progress in evolution grated against Darwin's non-directional evolutionary model. But there was nothing here that could not be found – to a greater or lesser extent – in Darwin, and Spencer did, contrary to what some believe, accept Darwin's fundamental point that evolution proceeded through destruction.[21] There was a mutual dependency, a grudging respect and a fundamental congruence of argument between the two men. Neither would have enjoyed the pre-eminence he did without the other. And Darwin's ideas would not have spread so quickly without the man he described as '[o]ur great philosopher'.[22] Both men offered permeable and evolving systems of thought which constantly overlapped and, taken in the round, taught the same lesson. When it came to writing about society the two were united in sprinkling their works with organic analogies and metaphors, and seeking the antecedents of human faculties, behaviour and history in the animal kingdom. And it was these 'social Darwinist' lessons that socialism learnt.

The social organism

The idea of an analogy between the body and society is as old as philosophy itself. Spencer's contribution, in his 1860 essay 'The social organism', was to reassert the analogy at a time when mechanistic metaphors dominated political economy, and to relentlessly explore the correspondence in madly precise detail. In the process of demonstrating his point that 'society is a growth and not a manufacture', Spencer found 'real parallelisms', which, he said, had been indiscernible to Plato, Hobbes and others without the benefit of physiological science.[23]

The absurd pleasure Spencer took in comparing the ruling classes, trading classes and masses to the mucous, vascular and serous systems of the liver-fluke, and in describing two French engineering écoles as 'a double gland . . . to secrete engineering faculty', suggest that he thought society was not only like an organism but really *was* an organism. This was a logical deduction from his synthetic philosophy. If man and society were part of the same cosmic evolutionary process, it was possible, indeed necessary, to understand both in terms of the same evolutionary laws. 'The human being,' as Spencer explained, was 'at once [both] the terminal problem of Biology and the initial factor of Sociology'. Societies, in common with individual organisms, commenced in small aggregations and insensibly augmented in mass, becoming increasingly complex and interdependent at the same time. Of course, there were differences, such as societies having no specific external form or continuous mass and all members being endowed with feeling, rather than this being restricted to a special tissue. Nonetheless, Spencer managed to qualify all the differences he noted and conclude that the '*principles* of organisation are the same, and the differences are simply differences of application'.[24]

The relevance of the organic analogy for Spencer's political position was that it allowed him to pitch the naturally evolving and self-governing social organism against a mechanistic state. On one level this was no more than a biological rendering of radicalism. But by making the case in biological terms Spencer opened his analysis up to a socialistic reinterpretation. One socialist who worked in this Spencerian framework and who directly influenced the imagery of London's *Iron Heel* was Laurence Gronlund. With his book *The Cooperative Commonwealth* (1884), this Danish émigré to the USA did more than any other single author to develop the socialistic potential in Spencer's thought.

Gronlund was unrestrained in his praise of 'the most profound of recent English philosophers'.[25] Spencer's 'speculations on the Social Organism' had, Gronlund argued, 'laid the foundations for constructive Socialism'. This was a new brand that Gronlund was keen to distinguish from the utopian and mechanistic varieties of Fourier and Saint Simon. The principal difference was that 'constructive Socialism' rejected the utopian strategy – which he disparaged as tearing down a building to erect a new edifice – and instead saw socialism as an evolutionary growth of the social organism. Gronlund was every bit as convinced as Spencer that society was 'a living Organism, differing from other organisms in no essential respects'. Indeed, by his own account, Gronlund insisted on this identity 'with even greater force than Spencer did'. In case there was any confusion, he emphasized that this was not to be understood as a metaphor. Society did not merely resemble an organism, but 'literally IS *an organism*, personal and territorial'.[26] All that differed from Spencer was the conclusion, for whereas Spencer, like Kropotkin, found in the evolution of the social organism evidence of the superfluity of the state, Gronlund read it as evidence of the need for an ever more socialistic state.

This involved developing Spencer's identification of the social organism in two ways: firstly, by equating Spencer's law of differentiation, which he said governed all organic change, with Kropotkin's law of mutual aid; and secondly, by making the state analogous to the governing function of the brain in the human organism. Neither involved a travesty of Spencer's principles and both positions were defensible on purely Spencerian grounds. Spencer's law of differentiation – that the law of evolution was to proceed from the homogeneous to the heterogeneous – was derived from the embryological works of the German physiologist Carl Ernst von Baer.[27] With a similar pattern discerned in the fossil record, where the more complex animals seemed to appear later, Spencer took this to be a law of all development, including social development, where an increased division of labour indicated a more highly evolved organism. For Spencer this confirmed his individualism – the more highly involved the society, the more differentiated the individuals. But equally it implied that the more differentiated the society, the more interdependent each of its constituents must be. This was close to what Kropotkin argued in *Mutual Aid*. And historically, a collectivist interpretation of the organic analogy, such as that found in the writings of Thomas Carlyle, was far more common than Spencer's individualistic one.[28]

Spencer, of course, maintained that his interpretation had 'modern science' on its side, but this, as many saw, was a doubtful claim.[29] His organic argument that the state should recede in size and scope with the advance of civilization was made possible by arbitrarily equating the state in advanced societies with the dispersed nervous systems of invertebrates, and by assuming that industrial activity was analogous to the internal alimentary activities of an organism – such as digestion and circulation – which functioned independently of the central regulatory mechanism. But this ignored Huxley's argument that with progression up the evolutionary ladder the brain became the 'sovereign power of the body', which it ruled with a 'rod of iron'.[30] This could easily be equated with the state in human society, rendering the growth of that institution both a natural and necessary part of the evolutionary process. And that is precisely what Gronlund did. In his hands, Spencer's organic analogy was developed into an organic argument for socialism.

Spencer's historical schema

George Bernard Shaw had copies of *The Cooperative Commonwealth* shipped to London in 1885 and one eager reader was Annie Besant. She developed Spencer's socialist potential in her pamphlets *Why I am a Socialist* (1886) and *The Evolution of Society* (1886). Besant praised Darwin, Huxley, Haeckel and Buchner for each illuminating 'the hidden recesses of Nature', but evolution, she said, was not confined to natural history. In the sphere of the mind and morals, 'Spencer was the great servant of Evolution, illuminating the previous darkness by lucid exposition and by pregnant suggestion'.[31]

In truth, even Besant's view of evolution in the natural world, in which 'the simple precedes the complex', and progress was 'a process of continued integrations and ever-increasing differentiations', owed as much to Spencer as the other authorities she cited. But when Besant came to ask 'Whither is society evolving?' she could not accept Spencer's individualistic conclusions.[32] Her own answer – 'towards that Golden Age which poets have chanted, which dreamers have visioned, which martyrs have died for' – hinted at a nostalgic hankering for a pre-industrial organic society.[33] But a Carlylean or an ethical case for socialism would not do. Besant's years as a secularist had convinced her of the need for scientific authority and she depended upon Spencer's methodology to 'grasp tendencies' that might indicate the future shape of society.[34] These, she claimed, pointed to an increasingly socialistic condition.

This conviction rested on some of the same grounds as Gronlund's but to a much greater extent Besant built on Spencer's account of historical evolution. As society was an organism this, of course, followed the same pattern as the rest of the organic world. The law of differentiation meant that societies evolved from the homogeneous to the heterogeneous, at the same time as the violence and struggle of natural selection gave way to more peaceful evolutionary mechanisms. But Spencer also had a more specific schema for the historical development of society. There was, he said, a general movement from militant to industrial societies. A militant society was characterized by violence and struggle. An industrial society was one in which peaceful and co-operative forms predominated. But neither was an ideal type. Rigid stages of historical development – such as those found in the four-stage theory of history of the Scottish Enlightenment – were too caesural for Spencer's organic turn of mind. Instead of sharply delineated phases, Spencer saw history as an almost imperceptible movement along a temporal continuum.

As they progressed, Spencer argued, societies became more differentiated and less violent. Militant societies were homogeneous, beginning in small, warlike tribes. In the course of their various wars and conquests, larger social units were created and these were less warlike. This was partly because there were fewer potential enemies and the risks of warfare were greater but also because the interdependence of industrial societies fostered a greater sociality and altruism, which militated against violent impulses.[35] In Besant these points were developed to argue that evolution was developing along a socialistic trajectory. Even more encouraging was Spencer's chapter on 'Sympathy' in his *Data of Ethics* (1879). In this he implied that egoism declined in the course of evolution, which – if taken with his suggestion in *Principles of Sociology* (1876–96) that evolutionary progress would take man to a future in which he looked for more than material aggrandizement and devoted his energies to 'higher activities' of an intellectual and moral character – was heavily socialistic in its implications.

Spencer was not a 'whatever is, is right' philosopher, or simply an unthinking apologist for *laissez-faire*. He believed that whatever was in line with evolution was right. And his vision of a higher evolutionary state could be used to justify socialism if one could demonstrate that this was compatible with evolutionary development. This was precisely what Besant attempted. In strictly Spencerian terms, she proceeded from the assumption that society was an organism; made a bio-social case that rested on a strict correspondence of biology and sociology; rooted her account of evolution in the law of differentiation; and saw society's evolutionary progress as being from militant to industrial society. An organism, she argued, could survive and prosper only through the further integration of its constituent cells and it was amongst those social animals 'who – like the bees and ants – have carried very far the subordination of the unit to the social organism that the most successful communities are found'.[36]

With the socialistic tendency of evolution asserted, she proceeded to make an evolutionary case for interventionist law making. This was justified in terms of the need to protect evolution. Egoism and individualism, she argued, were 'anti-social tendencies' that endangered the health of the whole organism and threatened to put evolution into reverse.[37] The evidence for this could be seen in the pollution of the rivers and the air and could be read in the stunted development of the short, bowed and pallid factory worker. Both were the result of greedy self-seeking. Legislative intervention, therefore, such as factory and sanitary Acts, was 'immediately necessary, in order that the environment may be changed sufficiently for the higher development of healthier organisms'.[38] The movement from militancy to industrialism, Besant concluded, was a movement from 'individualistic anarchy to associated order; from universal unrestricted competition to competition regulated and restrained by law, and even to partial co-operation in lieu thereof'.[39] Thus the tendencies one could grasp in legislation such as the Land Acts, Ground Game Acts, Education Acts, Shipping Acts, Employers' Liability Acts, and Artisans' Dwellings Acts pointed to an increasingly socialistic future.[40]

Conclusion

The evolutionary case for socialism, found in Besant, Gronlund and London, demanded something more than a demonstration that the tendency of organic change was to become steadily more socialistic. For political activity to be necessary and beneficial it also required that the evolutionary process be placed in danger. The ultra-optimistic early writings of Spencer gave little encouragement to such a view. But towards the end of his life Spencer's thought took a darker turn. As the fear of degeneration gripped the European mind, Spencer increasingly dwelt on the signs of regression to a more militant society. Kropotkin had expressed the same fear in *Mutual Aid* when he argued that the modern state had stripped men of the habit of social co-operation. And this fear of evolutionary reverse was to become a staple element of socialist thought – though Spencer and Kropotkin's ascription of blame to the state would not catch on. For most socialists the state was needed to rescue the evolutionary process.

This analysis affected their self-image. Socialists came to see themselves as redeemers of an evolutionary process that had gone wrong. Thus to be a socialist was to be a scientist, as much as an activist – someone who could study and understand the social organism, rather than someone who sought to create something new of their own. The mechanistic and revolutionary character of early-nineteenth-century socialism, that is, was eschewed, in favour of an organic conception of incremental change. The task of socialism, after all, was not to institute a new society. No one could impose such a thing on the social organism. The task of socialism was simply to put society back on its path of natural evolutionary progression.

All this can be seen in London's *The Iron Heel*. The idea that evolution has gone wrong is obvious as the degenerate mob rampages through the streets of Chicago, while the notion that socialists will redeem the evolutionary process is confirmed in the description of Everhard's crucifixion. Most of all, *The Iron Heel* makes it clear that the French revolutionary model of pre-Darwinian socialists is no more. Some of London's readers have not understood this message.[41] But Trotsky and others who looked to London for a straightforward endorsement of revolutionary politics were bound to be disappointed. The action on the streets of Chicago was reactive and provoked by the iron heel. It was necessary to stop those who would frustrate evolution but it would not create a new society, fully formed. That, as the publication of Avis Cunningham's manuscript seven centuries into the future indicates, would only arise after a very long evolutionary gestation.

The time lag was more than the artistic 'hyperbole' that Trotsky claimed.[42] London was consciously extending the time-frame found in socialist utopian novels such as William Morris's *News from Nowhere* – set a mere hundred years in the future – to make the point that socialism is an organic development, not a mechanism made by man. In his essay 'Revolution', written at the same time as *The Iron Heel*, London explained that the socialist revolution was not like its French or American predecessors. It was not 'a flame of popular discontent, arising in a day and dying down in a day'.[43] It was not incendiarism or insurrection. Nor would it be made by the immiserated masses. The socialist revolution was an organic process that accompanied the growth of civilization, just as it was an intellectual movement 'in line with social evolution'.[44] It could not be made. But equally it could not be stopped. At least, not for long.[45] Ultimately the iron heel would succumb and society would be restored to its path of evolutionary progression to an ever more socialistic state. As we shall see over the succeeding chapters, these incrementalist ideas dominated the late-nineteenth- and early-twentieth-century left.

5

Ramsay MacDonald: Ideologist of Evolutionary Socialism

'Socialism', declared James Ramsay MacDonald (1866–1937) 'is naught but Darwinism'.[1] MacDonald's contribution as an ideologist of socialism has suffered both from the bitterness that still lingers from his betrayal of 1931 and from the fact that his socialism was so firmly rooted in the concepts and language of evolutionary science.[2] His first theoretical work, *Socialism and Society* (1905), was also British socialism's first book-length attempt to unite socialism explicitly with Darwinism. The text was laced with analogies and metaphors drawn from biology and all of MacDonald's subsequent works were suffused with the same organic imagery and evolutionary language.

According to his friend and first biographer, Mary Agnes Hamilton, the importance of science to MacDonald's socialism could not be overstated. In her view it was MacDonald's scientific studies that distinguished his socialism from 'Liberalism on the one hand, on the other Marxism in its dogmatic form'.[3] But his scientific pretensions have left post-war historians and commentators embarrassed and nonplussed in equal measure. Some have mocked, others have shaken their heads in incomprehension, while Rodney Barker counselled his colleagues to work from the assumption that MacDonald did not actually mean what he wrote.[4] It was advice that Pat Thane obviously took to heart, ignoring MacDonald's claims to have rooted his socialism in Darwinism and instead attributing them to the more obviously 'political' Idealist influence of the Rainbow Circle.[5]

The open disdain with which historians treat MacDonald's own words is not simply due to a residual repugnance at his actions in forming the National Government. It also reflects a deep-seated suspicion of the role of scientific ideas on the left. Even David Marquand's momentous official biography treated MacDonald's interest in science as a youthful indiscretion, or a slightly embarrassing hobby that occasionally infected the language of the serious business of politics. Thus biologists such as Darwin, Huxley and Wallace – whom MacDonald himself judged among his most important intellectual inspirations – do not merit a mention in Marquand's 24-page index.[6]

Yet from his earliest writings MacDonald self-consciously developed a 'biological view' of socialism. Science was MacDonald's 'first love', and there was no division between his science and his politics: 'His conception of Socialism was biological; his biology was socialistic.'[7] According to MacDonald, the 'Socialist method' was the 'scientific method' and the 'Darwinian method', all three terms being synonymous. Geology, biology and socialism, he argued, were a trinity, three stages in one movement. Just as Charles Lyell's geological researches had paved the way for Darwin's advances in biology, so Darwinism, in turn, paved the way for socialism.[8]

This analysis mirrored MacDonald's own intellectual journey from an early interest in geology, through Darwinism and into socialism. However, to present this as a sequence is to miss the crucial point: MacDonald did not move *from* science *to* socialism. His interest in both subjects developed together; they fed off each other and became inextricably intertwined in his mind. He was not a scientist and then a socialist; he became both at the same time. Between the ages of seventeen – when MacDonald established the 'Lossiemouth Field Club' to undertake geological studies of the local area – and twenty-two – when a breakdown brought on by overwork robbed him of his long-cherished ambition of a scientific scholarship at the South Kensington Museum – MacDonald's science and socialism grew side by side.[9] In his private notebooks from the 1880s and 1890s politics and science jostle for space, cheek by jowl. Notes on organic chemistry nestle next to considerations on Church disestablishment, and notes on experimental physics are cut short by musings on the Poor Law.[10] When in Bristol, MacDonald's activities for the Social Democratic Federation (SDF) were curtailed by his undertaking a study of the Bristol geology. When he first came to London, meetings with Fabians were interspersed with evenings at the Birkbeck Institute and lunch-hours in the Guildhall Library, where his reading comprised articles on both geology and political economy.[11]

Nor was this merely a feature of his early years, a youthful hobby that was later cast aside. Even when in government in the 1920s and 1930s, MacDonald was still collecting scientific periodicals and lectures.[12] Indeed, at least up until 1914, his writings and speeches grew more, not less, explicit in linking Darwinian biology and socialism. Just as his first political lecture, 'Malthusianism versus Socialism', delivered to the Bristol branch of the SDF in 1885, has been described as lying on 'the borderland between politics and science', so the same could be said for all his major works in the Edwardian period.[13]

In addition, MacDonald used his editorship of the *Socialist Review*, the theoretical journal of the Independent Labour Party (ILP), to convince other socialists of the need to frame socialism in Darwinian terms by publishing many science-related articles. That MacDonald kept abreast of the latest developments in biology can be seen in the book notices he included in his 'Outlook' editorials. It was important, he commented when reviewing Professor Punnett's study of *Mendelism*, that '[s]ocialists who are anxious to test their faith by the most recent knowledge and to express it in accurate language, should keep their eye upon the newest biological literature'.[14] This was not least because such literature often provided the strongest arguments for socialism. Thus when noticing J. Arthur Thomson's *Darwinism and Human Life*, MacDonald told his readers that '[t]he final chapter in the book: *Selection, Organic and Social*, might well be issued as a Socialist pamphlet, though the definite idea of Socialism is not to be found in it.'[15] From this it ought to be clear just how intertwined were MacDonald's scientific and socialist thought.

MacDonald's Darwinism

His socialism was Darwinian and his Darwinism was socialist, although most historians have focused on the latter and ignored the former, taking MacDonald's idiosyncratic interpretation of Darwinism as evidence of a purely instrumental use of science.[16] But that is too simple. Certainly MacDonald interpreted Darwinism in a manner that would now be unacceptable. Socialism may have been 'naught but Darwinism' but only with a number of caveats – 'naught but Darwinism economised, made definite, become an intellectual policy, applied to the conditions of human society'.[17] Socialism was Darwinism 'applied to human society, with such modifications as are necessitated by the fact that they now relate to life which can consciously adapt itself to its circumstances and aid natural evolution by economising in experimental waste'.[18] That is, socialism was Darwinism without natural selection or overbreeding, without the struggle for existence or the survival of the fittest and with consciousness negating struggle, apart from that between ideas.

But in this MacDonald was no different from the other socialists we have studied. Indeed, he drew from most of them. From Kropotkin he learnt that the law of mutual aid was a necessary balance to the struggle for existence.[19] Following Gronlund and Besant, he looked for the socialist potential in Spencer. From Wallace, MacDonald learnt that man was uniquely able to defy natural selection by the power of mind. And from George, he drew the lesson that 'civilised man' had created a cultural space, beyond natural evolution and 'under the sway of the comparatively rapidly moving and acting human will'. Thus human evolution, said MacDonald, depended upon co-operation and intelligence, rather than the 'survival of the fittest' and death.[20]

MacDonald wore his intellectual debts as a badge of pride. His eclecticism was the product of his auto-didacticism and his belief that it was 'not the genius of man' but 'the growth of human knowledge' that had led to the establishment of evolutionary theory.[21] Both made it natural for him to draw freely from a range of authors. Darwin, Wallace and Spencer were accorded equal billing as the founders of evolutionary theory and MacDonald felt no need to reach a settled view of the mechanism through which evolution might work. 'Darwinian', in MacDonald's lexicon, was a distillation of the best in post-Darwinian biology.[22]

This loose use of the term, however, which he made interchangeable with 'biological' and 'evolutionary', should not be allowed to obscure the importance of MacDonald's scientific language to shaping and conditioning the content of his socialism. Darwinism, MacDonald said, offered a 'commanding standpoint from which to judge our Socialist proposals, a more accurate way of carrying them into effect, and a more scientific phraseology in which to express them'.[23] We cannot, therefore, sift out his scientific language as an extraneous element, inessential to his real political thought. MacDonald the scientist and MacDonald the socialist were not radically dichotomized. One was not prior to other. The concepts and language that MacDonald derived from evolutionary science were not a coincidental device distinct from his politics. His science did not determine his politics. But neither did his politics exist prior to his science. In MacDonald's thought the two fed off and reinforced each other.

MacDonald's understanding of Darwinism provided the discursive space in which he developed his socialism and this was to prove crucial to its content and identity. According to MacDonald, the character of a true socialist, as the example of Wallace demonstrated, was that of an evolutionary scientist. This severely limited his room for action. The 'supreme claim' of the socialist was that he alone understood the full meaning of the major tendencies in social evolution. And these were socialistic. But the best a socialist could do was to organize these tendencies. The social organism itself had to produce the forces that would drive society 'towards a higher social organisation', leaving the socialist creatively hamstrung.[24] A socialist could study, understand and occasionally guide the forces of social evolution, but he could never make or construct socialism.

With socialism defined as 'the method of evolution applied to society', the socialist method had to be 'organic and experimental' rather than 'architectural and dogmatic'.[25] 'Society', MacDonald warned, 'is not to be changed as we change the structure of a house; it evolves as a living thing does, changing with its organs and functions, the new forms never getting out of touch with the old.' Socialism was 'not a wiping-out but a transformation, not a re-creation but a fulfilment'. Social evolution entailed 'a progression of social stages which have proceeded and succeeded each other like the unfolding of life from amoeba to the mammal, or from the bud to the fruit'.[26]

This postponed 'the Socialist millennium till doomsday', and self-consciously marked MacDonald off from pre-biological socialists, such as Robert Owen, who through an 'error of temperament' had mistakenly attempted to build their new society by fiat.[27] This was of more than theoretical importance. The loss of faith in the socialists' ability to prescribe, which followed from MacDonald's Darwinian heuristic, helped foster the policy imprecision – 'a Turner landscape of beautiful colours and glorious indefiniteness', according to one critic – that characterized both MacDonald's theoretical writings and the legislative timidity that gripped him in office.[28] Both were symptomatic of a methodology that regarded any detailed plans for reaching the New Jerusalem, or any active interference to hasten its realization, as suspect.

The disabling effect of the argument that history was not a voluntary action and that socialism had to await the 'fullness of time' was counterbalanced by the certainty it engendered that evolution was on the socialists' side. Following Spencer's lead, MacDonald conceptualized the evolution of society within an historical schema, although in place of Spencer's two-stage model, MacDonald found societies evolving through three stages. The first two, the political age of the feudal past and the economic age of the capitalist present, were analogous to Spencer's militant and industrial societies. But no condition was eternal in an organism and the present was only 'a phase in the evolution of industrial organisation . . . not its final form'. Beyond capitalism lay 'the next stage in social growth', the character of which could be seen in the tendencies apparent in the existing social organism. Ever since the mid-nineteenth century, anarchic and mechanical individualism had been developing into a more organized form. And this increasingly socialistic state, said MacDonald, was rendering society 'more and more capable of expressing the moral constitution of man'.[29] Thus he called the next evolutionary stage, the moral stage.

As society evolved from one stage to the other, so it augmented individual ethical characteristics, such as altruism, and rose to a higher ethical plane. This equation of social evolution with the movement towards a higher morality owed something to Spencer and the phrenological assumption that, as man developed, his moral faculties assumed ascendancy. The proximate inspiration for MacDonald, however, was Wallace's argument that human evolution exhibited 'evidences of purpose and a foreordained end',[30] and entailed a progressive elevation of man's ethical nature.

Underlying MacDonald's analysis was the use of the organic analogy to understand society. In *Socialism and Society*, for example, MacDonald pursued this analogy in as mad and relentless a fashion as Spencer had. Indeed, as in Spencer, this was not really an analogy at all. Society, MacDonald admitted, was less rigid and more under the sway of the human will than the body, but its organization was literally 'biological'. 'The likeness between Society and an organism like the human body', MacDonald concluded, was 'complete in so far as Society is the total life from which the individual cells draw their life'.[31] The fact that individuals in society each enjoyed consciousness, whilst cells in an organism did not, or that society had no bodily form, were dismissed as less significant than the interdependence of the organs.

It was from this organic analogy that MacDonald began to distinguish the socialist position from that of liberalism. Against Spencer, and in common with Gronlund and Besant, he found a positive role for the state by equating it with 'the governing function' of an organism, thus negating Spencer's liberal objection that the state was a coercive force standing over society. MacDonald's organicism also led him to dispute liberal individualism. In *The New Charter* (1892), for example, MacDonald dismissed liberal individualism as anti-social, in much the same terms as Besant had. The liberals' 'atomic individualism', he argued, brought in its wake 'the desolating anarchy of *laissez-faire*'. The socialist, by contrast, understood that individuals lived as cells in a social body, and that true individuality could flourish only within a social environment. This required an 'organic individualism', by which he meant the differentiation of functions that occurred in organic evolution. And this would be achieved in 'an atmosphere of co-operation much better than in one of competition'. Thus the 'historic task' of socialism, to reconcile individual liberty with the needs of the social organism, involved the state actively providing opportunities for the exercise of individuality.[32]

The influence of sex on social progress[33]

A good example of how MacDonald developed a socialist argument within an organic and evolutionary discursive space can be found in an unpublished manuscript entitled 'The influence of sex on social progress'. This is the only surviving chapter of a projected volume entitled *The Socialist Polity*, which was never published. The Public Record Office has dated the manuscript as 1905, but this is clearly a mistake. The date scribbled in the margin at the end of the chapter is '12/5/90' and this is confirmed by the references to Queen Victoria, the New Fellowship, the Socialist League and Ibsen's *Doll's House*. The confusion may have arisen from MacDonald using the chapter when writing *Socialism and Society* in 1905. The date is important because it confirms that MacDonald was developing his socialism within an organic and evolutionary discursive space from the very beginning.

'The influence of sex on social progress' deals with the question of female emancipation and MacDonald self-consciously strives to build a distinctively socialist case for emancipation, which is rooted in evolutionary science and reacts against the classical liberalism of J. S. Mill's *The Subjection of Women* (1869). Mill had made his case in terms of an abstract individualism, which found no positive role for the state. Women were to be granted equality on the basis of a 'natural right' and the role of the state in policing marriage and preventing divorce was to be radically downgraded. MacDonald had no time for either position. To counter them he invoked Darwin, who had himself admonished Mill for making a natural rights case for equality, and Spencer, who had demonstrated that all social questions had to be 'explained, interpreted and classified in terms of evolutionary physiology'. But in MacDonald's hands the hereditary biological arguments, which Darwin and Spencer had used to provide an apologia for gender inequality, were transformed into evolutionary and anthropological argument for equality. The conclusion, that women should receive the vote and equal status with men, was the same as Mill's but the mode of reasoning was very different.

The crux of MacDonald's argument was that the stage of development of the social organism determined women's place in society. Women did not have a natural right to equality, as Mill asserted. Nor was it true, as Mill suggested, that male physical strength had hitherto excluded women from public life. In MacDonald's view, men could be absolved of all responsibility; they were 'as much the victim of circumstances as women'. A brief historical and anthropological survey testified that the social character of the successive epochs assigned women different roles. The question of female emancipation resolved itself into Spencer's point that what was right was whatever was in line with evolution.

Thus a natural rights argument gave way to an evolutionary and anthropological one. Early pleas for equality, such as Mary Wollstonecraft's *A Vindication of the Rights of Woman* (1792), had been 'produced out of time'. Women would only be admitted to the franchise once the social organism had evolved to the point at which female participation was beneficial. Just as the 'common people' had only been granted the vote 'after the state was sufficiently developed to require their aid', so the same was true of women. It was not men upon whom women's emancipation had been waiting, but 'the "fullness of time"'. Fortunately that time was now. The present social system was stunting female development to such an extent that 'The women of the nineteenth century are no more of the nineteenth century than are the Australian aborigines. They are in it but not of it.' Thus emancipation was justified in order to put society in line with its natural evolutionary advance.

In this way MacDonald gave a powerful endorsement to female emancipation. But the women's movement was guaranteed success only as part of a more general organic development. Just as Mill had argued that ending the subjection of women was merely an element in a broader historical movement from slavery and bondage to a free contractual society, so MacDonald found it but one element in 'a greater movement' to a more altruistic and moral society of socialism. There was, said MacDonald, 'a greater revolt' occurring against the 'world spirit of individualism and social masculinity', in which the women's movement was 'but a part'. This was understood in terms of Spencer's historical schema. Spencer's militant society was said to place the 'supremest social value' on war, competition and individual struggle – the 'distinctive features of masculinity'. The industrial society that succeeded it saw a decline of these masculine values, in favour of the feminine values of co-operation, altruism and morality. The next stage of socialism would require a further augmentation of these feminine values.

Far from sharing Spencer's dread of the female citizen, MacDonald felt the state had 'reached a period in its development' when male influence had to be supplemented and he hoped that the female franchise would bring a feminization of society in its wake: for example, introducing 'sympathy' 'as a legislative force'. Socialism demanded that the state developed the ethics of the family, in terms of mutual protection and care of the weak, and women were best able to promote this. Socialism was 'a new departure in thinking and in living' which required 'new help, . . . new hands, new inspiration'. The enfranchisement of women would 'not be merely an event in an epoch, but the beginning of a new epoch'.

This new epoch, however, would not signal any relaxation in the laws governing marriage; quite the reverse. By making the state the governing instrument in the social organism, MacDonald distinguished his position from that of both Mill and Spencer. He agreed with Mill that a true state of marriage could not exist until women were emancipated but he had no sympathy for the argument that divorce should be freely available. Divorce offended MacDonald's organic mindset – marriage by experiment was as ludicrous as surgery by experiment: one could not try out cutting off a leg. More importantly, Mill's case rested upon the assumption that divorce was purely a matter of individual concern. The same assumption was made in Ibsen's *Doll's House*, which in MacDonald's view epitomized 'selfish individualism'.

For MacDonald, marriage was an important social institution and its defence was vital to the health of the whole organism. Individual feelings of love, sympathy and affection were only the prelude to marriage. Following the legal ceremony, this 'spiritual union' became a matter of state concern. This is what Spencer had misunderstood when he claimed that the natural basis of marriage meant that over time the legal ceremony would die out, with love becoming a matter of individual arrangement – 'for a night, a year, or a life-time'. The primary task of the 'governing function' was to maintain the fabric of the social organism and this, MacDonald argued, justified maintaining marriages even against the will of the individuals involved. MacDonald was fully prepared to concede Bentham's argument in the *Civil Code* that to oppose divorce was to suggest that the state understood individuals better than they did themselves. The 'evil influences of an unfortunate marriage on the man, wife, children, and neighbours', he said, had to be balanced 'against the evil results of Society losing its interest in marriages'. The state 'actually [did] know how individual actions influence welfare better than the individuals concerned do'.

The influence of Idealism?

In the manuscript MacDonald made a passing reference to the Idealist philosopher T. H. Green. Green provided much of the intellectual inspiration for New Liberalism – the attempt of authors such as L. T. Hobhouse and J. A. Hobson to integrate organicism and a positive role for the state into liberal politics and philosophy. And some historians who have considered MacDonald's political philosophy have noted the congruence of his position with that of the New Liberals and concluded that MacDonald's organicism had the same Idealist root.[34]

This interpretation is understandable for those restricting themselves to overtly political influences, and MacDonald's membership of the Rainbow Circle, where he mixed with many leading Idealists and New Liberals, could be read as confirmation of a shared intellectual inheritance.[35] But our case suggests that the weight of evidence is clearly against this connection. Not only do MacDonald's own writings suggest *ad nauseam* that his political understanding, particularly of the role of the state, was built out of Darwinism rather than Idealism, but the chronology is clearly wrong. MacDonald, as we have seen, was a convinced organicist and evolutionist long before he joined the Rainbow Circle in the mid-1890s, and he remained one for the rest of his life. Stimulating though his time in the Circle may have been, one needs to keep a sense of perspective and bear in mind that MacDonald was an eclectic auto-didact, subject to a host of other influences. In particular, MacDonald was exposed not only to Darwin and Spencer, but also to socialist authors such as Gronlund and Besant, long before he would have encountered any representatives of the Idealist school. Texts such as *The Cooperative Commonwealth* and *The Evolution of Society* were both much closer in content to MacDonald's own writings than anything produced by the New Liberals, and would have been far more readily available to the young MacDonald than anything produced by Green and his Oxford colleagues.

Although historians have recently tried to resurrect British Idealism as a movement of major historical importance,[36] there must remain the suspicion that R. G. Collingwood's earlier characterization of the Idealists as a marginal group, writing impenetrable prose, and suspect even in Oxford, was more accurate.[37] Certainly the 'full connection between Idealism and socialism has yet to be fully spelled out',[38] but when it is it may be shown to be a tangential influence, which did no more than reinforce a pre-existent disposition to the use of an organic and evolutionary language among socialists. In any case, whatever the true extent of the influence of Idealism more generally, there is no escaping the fact that there were profound philosophical differences between the organicism of MacDonald and that of the New Liberals.

MacDonald's organicism was built upon a socialistic development of Spencer – whom he described as 'England's foremost philosopher' – and this was philosophically anathema to the Hegelian organicism that informed Idealism and underlay New Liberalism. This can best be illustrated by comparing MacDonald's approach with that of David Ritchie. Ritchie was unusual among the Oxford Idealists, not only because he was a socialist but also because he avoided the open disdain for Darwinism which, as Stefan Collini has pointed out, excluded the Idealists from influencing contemporary debates.[39] In Ritchie, therefore, if anywhere, we should be able to identify the possibility of Idealism seeping through into MacDonald's socialism. But on closer inspection we find that there were at least three important distinctions between MacDonald's and Ritchie's positions and these provide a template for understanding the distinctions between the organicism of the New Liberalism and that of Edwardian socialism.

Firstly, whereas MacDonald set great store by the introduction of a 'more scientific phraseology' into socialism, Ritchie found 'no special merit in the use of the word Darwinian' and felt historical and sociological works would be better off free of biological terms.[40] Thus while Spencer's linguistic excesses, in essays such as 'The social organism', encouraged MacDonald to be unrestrained in his own use of an organic and evolutionary language, the New Liberals tended to share Ritchie's circumspection. Ritchie was plainly irritated by Spencer's loose use of organicism, and willingness to apply an evolutionary language to sociology, and Hobson displayed a similar annoyance when reviewing MacDonald's *Socialism and Government* (1909). He could not, he declared, 'go as far as Mr MacDonald in his interpretation of the centralisation and specialisation involved in the organic treatment of society'.[41]

Secondly, MacDonald read in Spencer a deterministic understanding of evolution as a unilinear progress in which humans enjoyed only a minimal sphere for freedom and choice. Ritchie, by contrast, outlined a dialectical model of change, which implied both retrogressions and the ability of humans to make and remake societies through consciousness and rational choice.[42] Hobhouse's *Mind in Evolution* (1901), which was based on a close reading of Ritchie, posited a far greater freedom for consciousness and rational choice than the rigidly deterministic MacDonald ever allowed.[43]

Thirdly, while Ritchie sought to root his interpretation of Darwinism in Hegelianism, MacDonald saw the two systems as antagonistic and portrayed Darwinism as the antidote to Hegelianism. He could not share the relaxed attitude to Hegelianism that the New Liberals inherited from Idealism, because he was operating in a socialist context. And in socialist circles, Hegelianism equalled Marxism. Having done his utmost to throw revolutionism out of the socialist house, MacDonald was not inclined to let Hegelianism in through the back door of New Liberalism.

This last point is particularly worth emphasizing because it shows how deeply MacDonald's attitudes, unlike those of the New Liberals, were determined by debates *within* socialism and, in particular, how his thought was influenced by trends in European socialism. Marx was a *bête noire* for MacDonald because he was pre-Darwinian. Marx, said MacDonald, stood on 'the threshold of scientific sociology' but had been unable to cross it – 'Darwin had to contribute the work of his life to human knowledge before Socialism could be placed on a definitely scientific foundation.' The root of the mistakes of subsequent Marxists was to embrace a philosophy that rested on Hegel, who has been 'untrained in science'. The 'pre-biological' Marx needed to be brought up to date by integrating the insights of Darwinian science into socialism and eliminating any vestiges of Hegelianism, such as the dialectical method and the catastrophic and cataclysmic language that marred the *Communist Manifesto*.[44]

Having criticized Marx's dialectical schemes and his failure to guarantee that change equalled progress, MacDonald could hardly accept these same elements in Idealism.[45] Instead he developed his socialism out of a loosely defined, Spencerian-inspired Darwinism, which marked him off from Marxism on one side and liberalism, both old and new, on the other. Instead of chasing the will-o'-the-wisp of a New Liberal influence, historians would do better to locate MacDonald within the European Revisionist movement.

6

Marx and Engels

The relationship of Marx and Engels to Darwinism was different from that of every other socialist in this study. Considered purely on their own merits, Marx and Engels would not warrant a chapter in a short book seeking to show the extent to which post-1859 socialism was developed within a Darwinian discursive space. As made clear in the Introduction, Marxism was the one variety of pre-Darwinian socialism both to survive 1859 intact and to achieve a degree of political success. But Singer's critique of Marx, and his identification of the entire left with what he calls 'more broadly marxist (with a small 'm') thinking', makes Marx and Engels relevant to our story.[1]

The crux of Singer's case for a new 'Darwinian left' is that the left of the past 150 years has followed Marx's lead in 'know[ing] nothing at all about human nature'.[2] To back up this point Singer quotes from a part of Marx's Sixth Thesis on Feuerbach – 'the human essence is no abstraction in each single individual. In its reality it is the ensemble of the social relations' – and concludes that by this Marx means that if we change the latter we 'can *totally* change human nature'.[3] Then, telescoping a near century of historical development, and ignoring questions of historical contingency and specificity, Singer argues that this interpretation of human nature led to the horrors of Stalinism and adversely affected 'the thought of the entire left'.[4] The unspoken argument is that if only Marx and Stalin had been aware of the evolutionary psychology propagated at the London School of Economics, things might have turned out differently for the left. But it is Singer who labours under a misunderstanding, not Marx. His insertion of the word 'totally' in his description of what Marx meant in the Sixth Thesis is, as we shall see, tendentious. His claims that Marx and Engels were averse to Darwinism because of its Malthusian connection and distrustful of it as a product of bourgeois society are plain wrong.[5]

Few, however, have rushed to defend Marx and Engels. This is because Singer's criticisms touch a raw nerve on the left. For many reasons the dismantling of the myth around the *Capital* dedication was not the cue for a sustained exploration of Marx and Engels' intellectual relationship with Darwinism, or the natural sciences more generally. In fact Marx scholars, under the influence of Sartre and Marcuse, were at that time being drawn away from considering the role of science in Marxist thought and towards philosophy.[6] This was a movement that was given added impetus by the rediscovery of the 'early Marx'. Thus Singer's point that Marxism gives too little consideration to science hit upon a real failing in *contemporary* Marxism. Equally, his argument that Marxism is built around a rationalist denial of human nature, and the humanist desire to conquer nature, spoke to a real concern both within Marxism and on the post-1968 left more generally. The Green movement was conceived partly in reaction to the excessive rationalism and humanism of the former Soviet bloc, with the examples of the social engineering projects of Stalinist Russia and the environmental disasters of Soviet industrialization, vividly displayed at Chernobyl, providing the evidence.[7] From these criticisms it was but a small step to Singer's assertion that Marxism never really understood or accepted Darwinism. And the case is sealed with the *reductio ad absurdum* example, cited in almost every textbook, of the Soviet official, T. D. Lysenko, whose 1930s declaration of an ideological preference for Lamarckism over Darwinism had disastrous consequences for the Soviet wheat harvest.[8]

But Singer should not be allowed to proceed unchallenged. The case against Marx and Engels is sustained only through an equation of their position with the worst excesses of Stalinism. It is time the sins of the sons were stopped being visited upon the fathers. Marx and Engels have carried the incubus of Stalinism for too long. For good and bad 'Marxism exists in the nineteenth century like a fish in water' and only by returning Marxism to its natural habitat can we understand Marx and Engels as *nineteenth-century philosophers*.[9] From 1859 an essential element in their habitat was Darwinism. And their relationship with this new science was far more complex and nuanced than Singer allows. To study this we must first dispel the misunderstandings that cloud the stream.

Marx on nature

The most fundamental of these misunderstandings is the accusation that Marx denied the very concept of human nature. It is easy to see why commentators should have stumbled into this mistake. As a social revolutionary, Marx was concerned to emphasize the potential for human change. His focus had to be on the extent to which prevailing conditions and social relations were mutable and transient, not only in order to give hope to the oppressed that things could be other then they were, but also to expose the use of the category 'human nature' as an anti-socialist tool. To those political economists who considered capitalism the inevitable outcome of human nature, Marx had to reply that human nature itself could change. For example, in the *Communist Manifesto* those who invoked the 'laws of nature' were criticized for eternalizing contingent and historically specific social arrangements. And in *Capital*, Marx described how human needs were modified in the course of human history as a response to economic change – man 'acts upon external nature and changes it, and in this way he simultaneously changes his own nature'.

This emphasis on mutability led even some Marxists, such as Althusser, to conclude that Marx had denied that anything was eternal. But Marx did not seek to deny the concept of human nature, merely to limit its restrictive power. He wanted to demonstrate how our nature can change over time and how humans enjoy the freedom to evolve and develop in their social relationships. His ambition was to achieve a better balance between freedom and determinism, not to assert total freedom or make an outright denial of physical determinism. Marx's doctoral dissertation had addressed the contrast between the physical determinism of Democritus and the greater sphere of freedom allowed by Epicurus and it was just such a balance that Marx sought to assert in his writings on human nature. The import of his Sixth Thesis on Feuerbach was not that Feuerbach was wrong *tout court* but that Feuerbach referred too much to nature and too little to politics.[10] If Marx was guilty of anything, it was of 'bending the stick' too far against human nature, not of denying its existence.

Indeed, from the 1844 Paris manuscripts – in which 'human nature is profoundly discussed in a language of baffling complexity and continually shifting nuances' – through to *Capital*'s denunciation of the dehumanization and atomization of the labour process, Marx's writings were united by a critique of man's estrangement from his essential nature.[11] The underlying argument was that prevailing social relations needed to be replaced by alternative arrangements that were more consonant with the authentic parameters of the species. Capitalism, that is, was out of kilter with human nature or – to borrow Feuerbach's phrase, which Marx deployed in his essay 'On the Jewish Question' – was antagonistic to man's 'species-being'.[12] This argument can be found in all of Marx and Engels' early writings. In *The Holy Family* the proletariat's indignation with its own abasement was understood as 'an indignation to which it is necessarily driven by the contradiction between its human *nature* and its condition of life'. Equally, in *The German Ideology* the proletarian is described as one 'who is not in a position to satisfy even the needs that he has in common with all human beings', one whose 'position does not even allow him to satisfy the needs arising from his human nature'.[13] In Marx's later writings this hardened into the notion of alienation, but although the explicit references to human nature declined, the essence of the argument was unchanged. Alienation was a condition in which man's basic set of instincts, volitions and tendencies were contravened. Moreover, capitalism could not exist in perpetuity precisely because man was not infinitely malleable and, however much the 'ensemble of social relations' was changed, his essential nature would always remain antagonistic to the economic system.

Equally erroneous is the assertion that Marx and Engels evinced a humanistic contempt for nature that tainted the whole of the subsequent left. For a start, although Engels defined civilization as 'the gigantic control of nature' and Marx described Prometheus as 'the foremost saint and martyr in the philosopher's calendar', their desire to conquer nature was fairly muted by contemporary standards.[14] Whereas Wallace, for example, saw only the limitless advance of science and technology in the 'wonderful century',[15] Engels showed a greater ecological understanding by warning of human *hubris* provoking nature's *nemesis*:

> we by no means rule over nature like a conqueror over a foreign people, like somebody, thereby outside of nature – but that we, with flesh, blood, and brain, belong to nature, and exist in its midst, and that all our mastery of it consists in the fact that we have the advantage over all other creation of being able to know and correctly apply its laws.[16]

The example of Mesopotamia, where the destruction of the forests to obtain cultivatable ground had devastated the land by removing collecting centres and reservoirs, demonstrated that 'human conquests over nature' were often pyrrhic.[17]

This measured tone clearly separated Engels from what he elsewhere termed the 'senseless and anti-natural idea of a contradiction between mind and matter, man and nature, soul and body, . . . which obtained its highest elaboration in Christianity' and which Wallace had integrated into natural science.[18] Marx and Engels, by contrast, emphasized the continuity and interdependence of man and nature in a dialectical interaction. When Marx declared that 'nature, taken abstractly, for itself, *rigidly* separated from man, is *nothing* for man', he did not mean this to imply that man stood apart from, or separate to, nature.[19] His point was that abstract or deified concepts of nature, such as those found in Feuerbach, had no meaning: that man and nature were inevitably connected and existed only in conjunction. This led Marx and Engels to envisage something far more dialectical than a simplistic domination of man by nature, or of nature by man. Too many of their readers have only understood the latter process and denied the action of nature on man. But in *The German Ideology*, Marx and Engels condemned Christianity for regarding all determinism by nature as 'something foreign, a fetter, compulsion used against me'. A far more positive view of the influence of nature was expressed in Marx's assertion that communism offered 'the genuine solution of the antagonism between man and nature and between man and man' – not as a triumph of man over nature, but as a dialectical process in which labour 'humanises' nature and 'naturalises' humanity.[20]

Marx and Engels as scientists

Dispelling the misunderstandings surrounding Marx and Engels' attitude to nature and, more especially, human nature is only the first step towards probing their relationship with Darwinism. The second is to appreciate their abiding interest in the natural sciences, for although Engels demurred that he and Marx were able to 'keep up with the natural sciences only piecemeal, intermittently and sporadically', both were serious-minded enthusiasts.[21] Their correspondence frequently dwelt on the latest scientific views and their published works were peppered with scientific references. In Heyer's judgement '[i]t is hard to conceive of two persons, not scientists in the strict sense, who were more well read in the physical and natural sciences than were Marx and Engels',[22] although this rather misses the point that Marx and Engels were representatives of an age in which educated men switched easily between philosophy and science, without considering them strictly separate subjects.

In his youth Marx had been engrossed by Goethe's naturalism before discovering Hegel's philosophy. Whilst at Trier he worked under the renowned geologist Steininger and as a student in Berlin he attended lectures by the noted geologist and natural philosopher Heinrich Steffens.[23] Nor did this youthful passion for science dissipate, as Marx grew older. In London in the late 1840s he even developed a passion for phrenology. Although never sharing Wallace's wide-eyed commitment, he apparently 'believed in it to some extent' and Wilhelm Liebknecht recalled that upon their first meeting Marx subjected him to an examination by questions, 'but also with his fingers, making them dance over my skull in a connoisseur's style'.[24] This flirtation with what Engels dismissed as 'bumptious pseudo-science' soon passed and in later life Marx undertook intensive studies in the more reputable subjects of mathematics and astronomy, as well as occasionally dabbling in chemistry and biology.[25]

His broad interest in natural science meant that Marx encountered Darwin before most, and he was initially impressed. Marx first read the *Origin* in December 1860, just a year after it was published, re-read it in 1862, and may have attended lectures by Huxley in the same year. Liebknecht claimed that Marx's circle spoke of little else when the *Origin* was first published.[26] It was during this period that Marx famously wrote to Engels that the *Origin* 'contains the basis in natural history of our view'.[27] But his enthusiasm soon waned. By 1866 he was telling Engels that he preferred the work of the French naturalist Trémaux.[28] There is no evidence that Marx ever read the *Descent*, or any of Darwin's other publications. He did send Darwin a copy of *Capital* in 1873 – which to this day remains famously uncut in Down House – but one should not read too much into this. Marx also sent a copy to Herbert Spencer on the same day and few would attempt to use this as evidence of an intellectual affinity.[29]

The only personal contact of any note between Marx and Darwin was Edwin Ray Lankester, who acted as personal physician to Marx during his terminal illness and whose father had been a friend of Darwin. Thus with the shattering of the myth of the *Capital* dedication, the idea of an intellectual relationship between Marx and Darwin hangs by the thinnest of threads, two footnotes in *Capital* and a few scattered remarks in personal correspondence.[30] No wonder Marx scholars generally meet mention of Marx's indisputable interest in science with an indifferent shrug or an exasperated complaint about the difficulties of delineating any specific Darwinian influence.[31]

Engels receives a slightly different treatment. His retirement from business was the cue for an eight-year 'moulting' in mathematics and the natural sciences, as he attempted to bring his knowledge completely up to date, in preparation for a detailed work on the philosophy of science.[32] Two incomplete and rather unsatisfactory works resulted. The first, *Anti-Dühring* (1878), was a polemic against a German socialist contemporary who had assailed Darwinism as '*a piece of brutality directed against humanity*'.[33] Engels compared his task in this book to the necessity of sinking one's teeth into a 'sour apple'.[34] Dühring had to be answered but it was no fun for either author or reader. The book's chief virtues were that it demonstrated both the range of Engels' scientific interests and his sympathetic attitude to Darwinism. Of more value, but equally frustrating, are the unfinished manuscripts that make up *The Dialectics of Nature*. Engels' responsibility for editing and publishing Marx's unfinished volumes of *Capital* meant that the *Dialectics* remained incomplete at the time of his death.

Both *Anti-Dühring* and the *Dialectics* have been heavily criticized for entertaining a metaphysical conception of natural laws at odds with Marx,[35] just as Engels' interest in Darwinism has often been cited as evidence of how far he strayed from the philosophy of his friend and collaborator.[36] Against this, however, both works have won praise for demonstrating a sound grasp of the most up-to-date scientific knowledge available, including a detailed and profound appreciation of the details of Darwinism.[37]

But before considering Engels' use of Darwinism, and Marx's relative lack of interest, it remains to clear two further misunderstandings that Singer has helped to perpetuate. The first is that their anti-Malthusianism led Marx and Engels to reject Darwinism out-of-hand; the second, that their materialist interpretation of history led them to deny the legitimacy of any science produced by bourgeois society.[38] Both are caricatures of what Marx and Engels actually wrote. On the first point, both men were certainly opponents of Malthus. Marx echoed Cobbett in his virulent denunciation of 'Parson' Malthus, and Engels described Malthusianism as the 'crudest and most barbarous theory that ever existed'.[39] And equally both chided Darwin for his

Malthusianism. But they did not reject Darwinism on grounds of its Malthusian origins. Quite the opposite, Malthusianism ironically provided their principal point of agreement with Darwin. For both, the major achievement of the *Origin* was the deathblow Darwin dealt to teleology and 'the metaphysical conception of Nature'.[40] This achievement, which Marx picked out in *Capital* and which Engels highlighted in *Socialism: Utopian and scientific*, was rooted directly in Darwin's application of the Malthusian struggle to the natural world. This spelt the end of the sentimental deification of nature that Darwin objected to in Paley and Marx revolted against in Feuerbach. Marx and Engels had not weakened in their opposition to Malthus but recognized that Darwin's Malthusianism led him to a view of nature in accordance with their own. As Engels put it, '[h]owever great the blunder made by Darwin in accepting the Malthusian theory so naively and uncritically . . . no Malthusian spectacles are required to perceive the struggle for existence in nature'. The Malthusianism at the heart of Darwinism did not, therefore, invalidate Darwin's conclusions. In Engels' view, 'just as the law of wages has maintained its validity even after the Malthusian arguments on which Ricardo based it have long been consigned to oblivion, so likewise the struggle for existence can take place in nature, even without any Malthusian interpretation'.[41]

Science was thus accorded a status above and beyond the ideological function that it was said to serve, and even such an obviously ideological discipline as political economy was seen to have an inherent value. This was important because it meant that Marx and Engels' materialism was not necessarily a barrier to their appreciation of Darwinism. It is true that Marx embraced an interpretation of history which presented ideas and categories, including scientific ideas and categories, as '*historical and transitory products*', suitable for the prevailing relations of production, and that this threw the notion of eternal scientific truth into doubt.[42] It did not, however, lead to the kind of radical epistemological relativism we find in Foucault, which finds no truth or objectivity in science, only winners and losers and power. Marx combined his recognition of the underlying material influences on the production of scientific knowledge with a profound admiration for the work of individual scientists, including Darwin. Equally, a materialist framework did not deprive scientific study of its own internal dynamics of development. For example, in his essay on the rise of natural science, Engels explained the failure of Kant and Lyell to develop their respective theories of mutability into a general evolutionary view in purely internalist terms. It was not economic determinism or self-interest that hindered them but the force of tradition within natural science and the division of labour among natural scientists.[43] Marx and Engels, it goes without saying, would not have devoted so much time and energy to the study of science simply to write it off as the ideological offshoot of material interests.

Engels and Darwinism

Engels' use of Darwinism is interesting on three levels. Firstly, it demonstrates that it was possible to be an anti-Malthusian and a materialist yet still take Darwin seriously. Secondly, Engels' scientific interests are often cited as evidence of his divergence from Marx. Thirdly, Engels offers a striking point of contrast with the other socialist interpreters of Darwin. All three themes are amply demonstrated in his fascinating essay 'The role of labour in the transition from ape to man', which appeared in his *Dialectics of Nature*. Singer's view of the essay as a 'Lamarckian lapse' and evidence that Engels had 'not understood Darwin properly' only confirms his own ignorance of what Darwinism actually meant to Darwin and to the late nineteenth century more generally.[44]

Engels' readiness and ability to contemplate evolutionary mechanisms other than natural selection, including Lamarckism, sexual selection and Haeckel's models of 'adaptation' and 'heredity', are in fact proof that his understanding was at the cutting edge of contemporary science and, to a large extent, in line with that of Darwin himself. The law of correlated growth, for example, which lies at the heart of Engels' essay, was taken directly from Darwin and applied in an innovative fashion to the relationship between the hand and the brain in human evolution. But his argument that manual dexterity and mental capacity developed in conjunction was consistent with both Marxism and Darwinism[45] – the former by rooting human history in a materialistic account of the development of man's labour capacity, the latter by explaining the evolution of the human brain in terms that did not rely upon any special infusions or breaks in the evolutionary continuum. This is significant, for while Engels had a profound appreciation of Darwinism – he was prepared to utilize some Darwinian concepts – and reached some similar conclusions, these were always arrived at from a prior philosophical position.

In this sense 'The role of labour in the transition from ape to man' is misleading, not because it is a poor application of Darwinism but because it was exceptional to find Engels deploying an overtly Darwinian concept. His use of the law of correlated growth had enabled him to circumvent the problem of 'ultra-Darwinism' that Wallace had identified, of explaining man's mental capacities in terms of a strict natural selection. If humans had evolved mental capacities beyond the demands of utility, then this could be explained by an unintended correlation between growth of the brain and an augmentation in the potential skills and output of the human hand. Nor was this the only topic on which Engels took Darwin's side against his fellow socialist. Equally, in relation to Wallace's assertion of ruptures in the continuity of nature, analogous to the division between organic and inorganic matter, Engels found himself with Darwin, denying 'fixed lines of demarcation' and making continuity the 'kernel' of his conception of nature.[46]

But Engels' position pre-dated Darwinism. In a neglected essay entitled 'Natural science in the spirit world', Engels explained that the root of his differences with Wallace lay in their respective philosophical attitudes to science. Wallace, according to Engels, was guilty of the shallow empiricism that had blighted English scientists from the days of Bacon and Newton. This had put Wallace on a 'most certain path from natural science to mysticism' and accounted for his credulity and *naiveté* in embracing spiritualism.[47] By applying the same standard of proof to zoological studies as to the testimony of spiritualist experiences, Wallace inevitably failed to find a decisive falsification – even if one exposed a charlatan, this could always be taken as proof of the veracity of more ingenious tricksters.[48]

In place of this naïve English empiricism – which 'spurns all theory and distrusts all thought' and had led Bacon to lose himself in alchemy, Newton to die expounding the Revelation of St John and Wallace to fall for spirit-rapping – Engels proposed a more Germanic solution, a sound philosophical understanding of nature.[49] This, he said, is what underlay his own science. This philosophy was rooted in Marx's reinterpretation of Hegelianism, and had been largely unaffected by the publication of the *Origin*. Thus although in each of his points of dispute with Wallace – the law of correlated growth, the continuity of nature, the disavowal of spiritualism – Engels found himself in agreement with Darwin, he had reached his conclusions by a very different intellectual route. And here we have the crux of the relationship between Engels and Darwin: a congruence of conclusions derived from a different chain of reasoning.

Darwinism did not cause divergence between Marx and Engels, because Engels never forsook his Hegelian heritage. He was simply more concerned than Marx to emphasize the important points of congruence between Marxism and Darwinism. The first was that both doctrines saw nature as incessantly evolving and ceaseless in its movement. Darwin's critique of the static Paleyean universe was echoed in Engels' complaint that the 'narrow metaphysical mode' begot by Bacon and Locke understood nature only in isolation and death, rather than life and motion.[50] In nature 'nothing is eternal but eternally changing', claimed Engels, and this applied to man as much as the natural world.[51] Thus the second point of congruence was a shared assumption of continuity in nature. The full import of Darwin's enigmatic end to the *Origin* – that 'light will be thrown on the origin of man and his history' – in establishing the unity of man and nature was more explicit in Marx and Engels. When Marx described nature as man's '*inorganic* body' he sought to emphasize that man and nature were one.[52] There were no special interventions or 'spiritual infusions'. Man and nature were directly conjoined and their existence was continuous. Nature was both the 'direct means of life' for man and 'the material, the object and the instrument of his life activity'.[53]

In other words, human life developed in a continual exchange with nature.

Thirdly, and paradoxically for two revolutionaries, Marx and Engels shared much of Darwin's evolutionary model of change in nature. Although Marx took from Hegel and Feuerbach a phenomenology of the natural, he remained silent on Hegel's broader philosophy of nature, especially the cataclysmic successions, aware that the nineteenth century had rendered it obsolete as science.[54] Engels, in particular, saw Darwin's notion of nature developing in time as an advance on Hegel's model of co-existence. But in case there was any doubt as to their order of precedence, Engels made it clear that Darwin's real importance lay in re-establishing the possibility of a dialectical understanding of nature – by dissolving rigidity and dissipating fixity in nature – and providing practical proof of Hegel's account of 'the inner connection between necessity and chance'.[55]

Human and animal society

A congruence of conclusions masked not only a very different chain of reasoning but also a fundamental antagonism of methodology. This was rooted in Marx's view of nature, not, as Singer believed, because Marx denied the concept of human nature, but because Marx and Engels held a dialectical view of the relationship between man and nature that was at odds with the simplistic and one-sided application of Darwinism to the study of man. Marx and Engels rejected the idea that the study of the animal kingdom could teach meaningful lessons about the nature of human society. Yet as Darwin's comment at the end of the *Origin* – 'light will be thrown on the origin of man and his history'[56] – indicated, this was the key to the application of Darwinism to the study of man. It was this view that Marx and Engels mocked as Darwin's 'bitter satire', in showing 'that free competition, the struggle for existence, which the economists celebrate as the highest historical achievement, is the normal state of the *animal kingdom*'.[57] A shared opposition to Wallace's crude idealism, therefore, which separated man from the rest of creation, masked two very different views of the positive relationship between man and nature. Whereas Darwinists dissolved man and society into nature, Marx and Engels had a more sophisticated position. Their starting point, recalled by Engels even in the 1880s, was Feuerbach's dictum that 'Man as he sprang originally from Nature was only a mere creature of nature, not man, Man is a product of man, of culture, of history.'[58] And, Marx might have added, so too is nature.

Instead of legitimating a transfer of the laws governing the animal kingdom to human society, Marx and Engels found the continuity of man and nature proof that natural laws were inappropriate for understanding society. Their point was that man existed within nature but by his very existence changed nature. Human purposes had rendered the operations of 'nature' no more 'natural' than the operations of the free market. Nature – as an abstract, non-human entity – had no meaning. Man was so powerful that he had distorted nature itself beyond all recognition. Even the senses of social man – his musical ear and eye for aesthetic beauty – were superior to those of non-social man, the product of his history rather than his nature.[59] Social history and biology, therefore, had been rendered one and the same, not because one subsumed the other, but because human history inaugurated a new phase in which 'natural history' became 'human history'. Natural science could provide only a pre-history of man as an animal; it held no insights into his human and historical existence. Human history was not natural history. 'History', Marx wrote, 'is the true natural history of man.'[60]

Even Engels – supposedly the more Darwinian of the two – found animal societies of strictly negative value in drawing conclusions about human societies. Specifically, there could be no transference of the concept of struggle in nature to the class struggle in history.[61] Socialists, such as Büchner, who sought to understand civil society in terms of nature, were condemned as contemptible purveyors of 'shallow nonsense'.[62] Nature could not provide an objective criterion by which to judge humans. This was the point Marx made when he referred to Darwin in *Capital*, and he even quoted Vico: 'human history differs from natural history in that we have made the former'.[63] That Engels accepted it too can be seen in his *Origin of the Family, Private Property, and the State* (1884). The title was suggestive of Darwin but the book asserted women's oppression as a problem of history, rather than of biology.

The key to human history – and the distinction between man and the other animals – lay in labour. There could be no legitimate transfer from the animal kingdom to human society because human labour was unique. Too few understood this. According to Engels, even the most materialistic of Darwinian natural scientists were liable to bouts of idealism and a tendency to credit the human mind with advances in civilization.[64] Marx and Engels, by contrast, understood the origin of man in purely materialistic terms. Labour in 'an exclusively human characteristic' was distinguished from labour at 'the animal level' both by the use of tools and by the fact that man, with a conscious purpose, could realize himself through labour. This had no parallel in even the most intricate labour of animals such as the bees, beavers or ants. An animal is only able to express its species nature 'one-sidedly', by producing for immediate needs and with the product belonging intrinsically to its physical body; a continuing replication of an unchanging standard of the species. Man, by contrast, produces without immediate need and 'freely confronts his product'. Whereas the animal 'merely *uses* external nature', man '*masters* it' and in the process transforms nature.[65] The dawn of human history was the death knell of social explanations rooted in natural history. Marx and Engels would not countenance drawing lessons about human society from the animal kingdom. Their use of Darwin, such as it was, was occasional, inconsequential and usually designed to bolster a common conclusion reached by a very different intellectual path.

Conclusion

Marx and Engels had a fully developed and coherent intellectual system prior to the publication of the *Origin*. They were of the last generation of socialists who could claim this. Thereafter socialism, by necessity and choice, was developed within a Darwinian discursive space. This applied to Marx and Engels' closest intellectual heirs, as much as to their Revisionist critics. We have already seen that Edward Aveling, the partner of Marx's favourite daughter, Eleanor, sought to unite Marxism and Darwinism. In the next chapter we will see that Karl Kautsky, keeper of Marx and Engels' literary inheritance, sought to provide Marxism with an ethical basis by building out from the theory of sociability found in Darwin's *Descent of Man*. It was a mode of reasoning that Marx and Engels would have found alien and unacceptable, but one which, as we saw in the Introduction, Engels had inadvertently helped to promote by his own use of Darwin's name to add legitimacy to Marx's arguments.

Engels must also bear a large burden of guilt for the late-nineteenth-century socialist obsession with defining socialism as 'scientific'. Nowhere is this responsibility clearer than in his little pamphlet, *Socialism: Utopian and scientific* – significantly translated into English in 1892 by Aveling. The pamphlet, which grew out of Engels' dispute with Dühring and enjoyed more translations than the *Manifesto*, became one of the most influential texts on the late-nineteenth-century left.[66] By instilling the very simple idea that socialism could and should be 'scientific', this pamphlet did more than any other single publication to encourage socialists to look to Darwinism for the foundation of their politics. This may have been what Aveling hoped, but it was not what Engels intended. He had been concerned to distinguish Marxism, as a materialist doctrine, from the abstract reasoning of Owenism, Saint-Simonianism and Fourierism.[67] But in doing so, Engels bequeathed to the left an abiding horror of pre-scientific thought – which, in truth, Marxism itself was – and a desire to define late-nineteenth-century socialism in opposition to its early-nineteenth-century variants. This was understandably read as the need to develop a socialism that was in line with the latest science, and that inevitably meant Darwinism. Utopianism thus became a pejorative term, against which socialists defined their own position. The 'new' socialism was called by turns 'scientific', 'modern' or 'constructive' but it was always ostensibly in line with, and developed out of, Darwinism.

7

The Revisionist Controversy

The extent to which Marxism became entwined with Darwinism can be seen in the Revisionist controversy that engulfed the international socialist movement at the turn of the century. German socialists had long been developing their politics within a Darwinian discursive space when, in late 1890s, a debate surfaced within the SPD concerning the relevance of Marxism to contemporary society. From the start this was something far more serious than an intra-party squabble, both because the SPD was the largest and most successful socialist party in the world – and therefore any controversy within it mattered to the whole of the socialist movement – and because the debate touched upon how socialists should interpret their creed in Darwinian terms.

The two principals were Karl Kautsky and Edward Bernstein. Kautsky was the 'Pope of Marxism', the official interpreter of Marx and Engels' legacy, and the most outstanding theoretician of his generation. Bernstein was the heretic, who had the temerity to suggest that some favourite Marxist shibboleths, including immiseration, the increasing concentration of capital and the progressive hardening of class divisions, had been disproved by historical experience. His conclusion, that these mistaken doctrines ought to be discarded and the socialist analysis 'revised' accordingly, earned his doctrine the pejorative name 'Revisionism' and secured him a position of infamy in the socialist pantheon. Kautsky emerged as the formal victor. Revisionism was rejected by overwhelming majorities at both the 1899 and 1903 SPD conferences and few within the Second International – the loose confederation of socialist parties established in 1889 – were prepared to identify openly with Bernstein.[1]

But as is so often the case, the apostate was defeated only for the heresy to flourish. The programme of the SPD was, in practical terms, thoroughly Revisionist and by 1914 Revisionism 'dominated all but a small part of European social democracy'.[2] This was as true of Britain as elsewhere. On the whole MacDonald thought Revisionism an 'unfortunate term' because of its negative connotations and association with revising a Marxist tradition which was, in any case, largely absent in Britain. But he was a close friend of Bernstein and was happy to identify his own approach as 'Revisionist' and state his aim of building 'a strong school of Revisionist Socialism' in Britain. This was because MacDonald understood the positive side to Revisionism and shared Bernstein's desire to change the language and mindset of the socialist movement – to make it more organic and evolutionary in its analysis and prescriptions. Marxism was to be revised but in a way that would make it more compatible with Darwinism.[3] And, as we have seen, this positive aspect of Revisionism coincided perfectly with MacDonald's own objectives.

Bernstein's evolutionary socialism

Born in the year that Darwin's *Origin of Species* was published, the scientifically trained Bernstein made an interesting apostate. As a young man he had been part of the intimate circle of young German socialists – along with Bebel and Kautsky – who surrounded the ageing Engels at a time when the old man's interest in Darwinism and all things scientific took a firm hold.[4] The crucial experience that sparked Bernstein's anti-Pauline conversion was his extended stays in London in the early 1890s. Moving in Fabian circles he – according to taste – either developed an Anglophilia or contracted the 'British liberal disease'.[5] What is not in dispute is that he abandoned the dialectical assumptions of Hegelianism and accepted a simpler, unilinear model of progress. The implications of this for his socialism were first revealed in a series of articles published in the SPD's theoretical journal *Die Neue Zeit* between November 1896 and the summer of 1899. By the time the final essay appeared, Bernstein's thoughts had been expanded into book form in *Die Voraussetzungen des Sozialismus und die Aufgaben der Sozialdemokratie* (1898). The book consisted, as the literal translation of the title – 'The presuppositions of socialism and the tasks of social democracy' – would suggest, of two main arguments. The first was to challenge those of Marx's 'presuppositions' that were no longer relevant. The second was to complement this with a positive programme for the future 'tasks of social democracy'. With class conflict eschewed, and the imminent collapse of capitalism disproved, these tasks resolved themselves into a gradualist extension of democracy and an incremental winning of ameliorative measures to alleviate poverty and improve welfare.

Bernstein's *coup de grâce* was to present this not so much as a 'triumphing over Marxism' but as 'a rejection of certain remains of Utopianism' which distorted the Marxist analysis.[6] In his view, Marxism was disabled by a dualism; it attempted to be anti-utopian and evolutionary even while it clung to the revolutionary language of Blanquism and a model of catastrophic change derived from the French Revolution. Marx and Engels, that is, had not gone far enough in their scientizing of socialism. Despite their critique of the utopian socialists, the *Communist Manifesto* had left 'a real residue of Utopianism in the Marxist system', which neither Marx nor Engels had been willing to eradicate fully.[7] The Marx of *Capital* had demonstrated a greater understanding of the evolutionary development of society but, at the point when the final aim of socialism entered into the question, became 'uncertain and unreliable', relapsing into his youthful Jacobinism.[8] Engels too was chastised for failing to abandon revolutionary and catastrophic models completely. In his prefaces to Marx's *Class War in France* and the *War of the Classes*, Engels had recognized that political surprises were increasingly rare, and that a longer than anticipated time was needed for the socialist transition, but he had failed to understand the implications of this for socialist activity.[9]

Bernstein, therefore, was able to present himself both as more 'scientific' than Marx and Engels and as their successor, who was simply 'developing the evolutionary principle' latent in their works.[10] This was disingenuous. Bernstein was doing something more than tidying up a few loose ends. By disavowing revolutionary and catastrophic models of change, and finding Marx's theoretic importance in his evolutionism, Bernstein was altering the very essence of Marxism. He snatched at the evolutionary passages of the mature Marx in order to wield them as a stick with which to beat the Blanquism out of the *Manifesto*-Marx. In the process, he stripped Marxism of its dialectical model of change and inserted a unilinear concept of progress that owed more to Spencer than it did to Hegel.

Essential to this was a linguistic change. Underlying *Die Voraussetzungen* was an argument concerning the language that socialists should use. As a neo-Kantian, the importance of language to Bernstein's thought cannot be overstated. And in his concluding chapter, 'Kant against cant', Bernstein issued strictures against the socialist tendency to deploy unreal manners of speech, or to cling thoughtlessly to speech patterns and phrases from a bygone era.[11] Modes of expression, Bernstein believed, conditioned systems of thought. This made it essential for socialism 'to emancipate itself from a phraseology which is actually outworn' – that mechanistic, utopian and revolutionary language which socialism had inherited from Jacobinism – and equally important to adopt, in its place, a more accurate language and imagery.[12] Thus Bernstein's unremitting use of an organic and evolutionary language throughout *Die Voraussetzungen* was neither accidental nor incidental. Capital 'evolved' and grew through an 'organic combination'; co-operative stores were 'an organism fit to perform its work and capable of a high degree of development'; trade unions grew from 'the most elementary organisms'; and society was 'no firm crystal, but an organism capable of change and constantly engaged in a process of change'.[13] So when MacDonald suggested *Evolutionary Socialism* as the title for the English translation, Bernstein – who had meticulously corrected the translator when he felt there was a more precise word or phrase with which to express his argument – immediately agreed it was 'just the thing'.[14] To complain that 'there is no evidence of Darwinism in this text, despite its title', therefore, is rather to miss the point.[15] Bernstein was Darwinian in the language he chose to express his socialism and this was as important as the details of his political message.

There was no division of language and analysis; in Bernstein each reinforced the other. His biological language was his politics. For example, Marx's model of an increasingly stratified two-class society was at odds with all known examples of organic development, in which organisms grew more, not less, complex. 'Far from society being simplified as to its divisions with earlier times', the middle class was not disappearing and society had become more 'graduated and differentiated both in respect of incomes and business activities'.[16] Social development, therefore, had followed the same evolutionary pattern of movement from homogeneity to heterogeneity that von Baer had identified in other organisms. Not even the proletariat had become that 'homogeneous mass, devoid in an equal degree of property, family, etc.' which the *Communist Manifesto* had predicted.[17] Rather, their division into trade unions highlighted the internal differentiation of the working class.

Moreover, by conceiving society as an organism, the conceptual space for class antagonism was narrowed. Different classes within the same organism could not, by definition, have fundamentally different interests. If criticisms of cxploitation and injustice were to be made in organic terms, they had to be expressed as a form of parasitism, rather than class antagonism. And this changed the historic task of the working class from the violent destruction of capitalism into the absorption of 'the parasitic elements of the social body'.[18] This was also consistent with the understanding of change as passive growth, rather than revolutionary action, which Bernstein had imbued from biology. Socialism was not to be realized by the triumph of one class over another but as an accentuation of natural trends towards partnership and association within the social organism. There was to be no sudden change or the imposition of a plan, simply small, incremental steps – factory legislation, the growth of local government and the freeing of the trade unions – whose cumulative effect was to bring a new society into being.

This language determined the very character of Revisionism. It was Bernstein's Darwinism that separated the immediate tasks of socialists from the final end of socialism. With utopian strategies condemned as misguided attempts to foreshorten the evolutionary process, all that socialists could do was diligently identify, guide and nurture those forces within the social organism, which would lead the capitalist caterpillar to transform into the socialist butterfly. As with MacDonald, Bernstein's acceptance of 'organic evolutionism' meant that socialism ceased to be an absolute idea and became merely the expression of a tendency.

The end-point remained a mystery, as in any other process of evolution. Indeed, in the evolution of an organism there was no beginning and no end. This is what Bernstein meant when he declared that '[u]nable to believe in finalities at all, I cannot believe in a final aim of socialism'.[19] No one could foresee or foreshorten the course of evolution. Thus there was no point in concerning oneself with the final stage of evolutionary development. The task of the socialist was rather to busy himself with 'the duties of the present and nearest future' and to undertake the 'every-day work of the socialist party'.[20] This entailed building upon liberalism – to which socialism was the 'legitimate heir' – and accentuating all those tendencies within the social organism that encouraged partnership, association and co-operation.[21] This was more fruitful, and more certain of success, than idly speculating as to the final form that socialism might take. The socialist could carry out these everyday tasks safe in the certain knowledge that he was easing the path of inevitable evolution. Thus the essence of the implications of an organic and evolutionary language for socialism and the essence of Revisionism were the same: the movement is everything, the final aim is nothing.

Karl Kautsky

Bernstein's aphorism reflected the real spirit of German social democracy but with 'the final goal', as Trotsky noted, 'kept in Kautsky's department'. The everyday activity of the SPD was determinedly Revisionist but the Marxist notion of the proletarian revolution remained 'a theoretical generalisation and a historical perspective' in the works of 'the most outstanding theoretician of the Second International'.[22] Born in Prague in 1854, Kautsky had grown to political maturity during the reformist epoch between the rout of the Paris Commune and the first Russian revolution of 1905. Unlike Bernstein, however, he found a positive value in the term 'revolution'.

As the editor of *Die Neue Zeit* from 1883 to 1917, Kautsky had a greater influence than any other individual on the development of European socialist thought in the period before the First World War and was perhaps the only theoretician capable of convincingly rebutting Bernstein. But Kautsky was 'tardy and hesitant' in his answer.[23] Considerations of personal friendship, the desire for party unity and a distaste for polemics allowed Parvus, Rosa Luxemburg and Plekhanov each to respond to Bernstein before Kautsky's rather pedestrian *Bernstein und das sozialdemokratische Programm* (1899) appeared. Even then Kautsky restricted himself to a detailed and dry refutation of Bernstein's statistics, without engaging in the broader philosophical questions that *Die Voraussetzungen* had raised. Only ten years later, with the publication of *The Road to*

Power: Political reflections on growing into the revolution (1909), did the gloves really come off.

Part of Kautsky's problem was that Darwinism was as central to his thought as it was to Bernstein's. This meant that methodologically and temperamentally they shared the same preoccupations and assumptions. As a young man Kautsky had been 'raised on a steady diet of positivism and evolutionary materialism', and his earliest writings had been concerned with developing the writings of his favourite thinkers – Darwin, Haeckel and Büchner – into the service of socialism.[24] The most influential Marxist text in Kautsky's development was not *Capital*, or even the *Communist Manifesto*, it was Engels' *Anti-Dühring*, and throughout his life, especially in his magnum opus, *Die materialistische Geschichtsauffassung* (1933), Kautsky returned to Darwinism to lay the foundation for his socialist thought.

Equally, Kautsky shared with Bernstein an indomitable optimism, which in Kautsky's case survived even the rise of the Nazis, and an abiding horror of utopianism. Kautsky's works abound with references to 'necessity' and 'inevitability' but have little to say about 'making' or 'creating'. During his long editorship of *Die Neue Zeit*, only one article dealing with future society appeared and that was Kautsky's discussion of past millennial societies.[25] Even his study of Thomas More's *Utopia* was an extended reassertion of Engel's distinction between 'scientific' and 'utopian' socialism, concluding that More 'could not help being an Utopist'; his 'tragedy' was to have divined a problem before the material conditions were available for its solution.[26] Moreover, Kautsky's writings were as rife as Bernstein's with organic analogies and references to the social organism. He struck an almost Spencerian note in objecting to Kant's depiction of human society as equivalent to trees grouped into a wood:

> Kant has no idea that society is more than a collection of individuals living together, that it is an organism, the cells of which, the individuals do not simply interact, but by virtue of the division of labour have definite forms of co-operation, which depend not on their good or evil will, but on the nature of the productive forces.[27]

Where Kautsky differed from Bernstein was in apparently reconciling this organic language, and his Darwinian framework, with the classical Marxist case for revolution and class struggle. He did this as part of his broader critique of the neo-Kantianism on which Bernstein's analysis rested – a task made all the more necessary because the appeal of neo-Kantianism on the German left was not limited to the Revisionists.[28] Otto Bauer, for example, also looked to Kant for the system of ethics that they felt Marx's economic materialism lacked.

Kautsky's attempt to counter this tendency, in his *Ethik und materialistische Geschichtsauffassung* (1906), led him to Darwin's *Descent* and

Kropotkin's *Mutual Aid* for an understanding of morality that was rooted in man's animal origins. Moral and immoral, Kautsky argued, were not absolutes in the Kantian sense but relative concepts that varied according to the level of material development. 'The only form of absolute immorality would be the absence of those social instincts and virtues which man has inherited from social animals.'[29] Any other forms of morality were 'conditioned by transformations which society undergoes as it is driven forward by technical innovation'.[30] A moral ideal, therefore, was a very shaky basis on which to erect the case for socialism. Indeed, this was to repeat the mistakes of the utopian socialists. The triumph of scientific socialism – 'the scientific investigation of the laws of development and motion of the social organism for the purpose of ascertaining the necessary trends and goals of the proletarian class struggle' – was a far surer foundation.[31] According to Kautsky, the resultant transformation of socialism from a moral ideal into an economic movement, rooted in 'the given material foundations', did not strip socialism of its greatness.[32] It merely recognized that, while moral indignation could act to organize and inspire, only scientific study of the conditions and development of the social organism could provide directives for action.

Revolution and class struggle

From this defence of scientific socialism, Kautsky proceeded to make the case for retaining the notions of revolution and class struggle in Marxist thought. He did so, however, only by changing the philosophical basis on which Marxism rested. He was almost as guilty as Bernstein was of eschewing Hegelianism in favour of a naturalistic methodology drawn from Darwinism. On the question of revolution, for example, Kautsky's understanding was as Darwinian as Bernstein's but with a different emphasis. The subtitle of *The Road to Power*, 'political reflections on growing into the revolution', signalled Kautsky's unease with the Revisionist assumption that there could be an unproblematic evolutionary *growth* into socialism that was barely discernible, though ultimately profound.

Kautsky was prepared to admit that this assumption contained 'a core of unquestionable reality'. After all, Marx and Engels themselves, said Kautsky, had noted society's evolutionary progression towards an ever more socialistic state 'and demonstrated that it unfolded as though a matter of natural law'.[33] But the prospect of this continuing, without catastrophe, to the point at which socialism was achieved rested on two misunderstandings of the evolutionary process.[34] The first was that society was not one unified organism, as Bernstein assumed, but two antagonistic elements. Capital and labour evolved in parallel – capital becoming increasingly concentrated and the proletariat increasingly organized. In the process, their antagonism intensified and this almost guaranteed that the transition would not be peaceful. A great political convulsion – a revolution – would first have to occur. Bernstein's second mistake was to forget the centrality of 'struggle' to the Darwinian system. It was only 'in and through struggle' that the future society – analogous to a new species – could be formed.[35] Thus Kautsky made 'struggle' – meaning class struggle – the defining feature of the work of social democrats. But he removed Darwin's uncertainty about the outcome of any struggle, by introducing the reassurance that the proletariat must emerge victorious, due to the fact that evolution had made them indispensable to the life of the social organism and rendered capital superfluous.[36]

Kautsky's retention of the concept of revolution was achieved only by redefining the term. His was a revolution in what he called the politico-economic sense', not 'in the criminal sense' that Bernstein had understood.[37] This revolution was both the denouement of a long evolutionary advance – the decisive moment at which the quantitative weight of reform signalled a qualitative change in society – and, as in London's *Iron Heel*, a negative necessity imposed by the capitalists' resistance to the evolutionary process. This moment was denoted by the term 'revolution', as this helped highlight the necessity for conquering political power – the final objective of any proletarian party. But in practical terms, Kautsky was no more of a revolutionary than Bernstein. He agreed with Bernstein that the revolution was not to be made by a conspiracy, a *coup d'état* or a small minority – that would be utopian – and defined the SPD as 'a revolutionary party, but not a party that makes revolution'.[38] Nor would the revolution necessarily entail bloodshed, arson, anarchy or pillage. Theoretically, insurrection could be deployed as a tactic for seizing political power, though only once the pre-conditions for socialism, 'which are formed only gradually', were in place.[39] But the thrust of Kautsky's argument was that the revolution, which would last 'perhaps a generation', was itself a gradual, largely non-violent process: 'a significant political crisis which intensifies and quickens the pulse of the political life of the nation in contrast to counter-revolutionary crises which have the opposite effect'.[40] For Bernstein this redefinition was a good example of exactly the sort of cant – an evolutionary revolution! – that he was seeking to cut through.

Kautsky's account of the class struggle that made this revolution inevitable was even more explicitly dependent upon Darwinism. It began in his Darwinian explanation of morals. The assertion that material conditions governed morality raised questions about the extent of human free will, which Kautsky did not shy away from. Economic necessity, he argued, did not mean the absence of will in human society. But while Kautsky was prepared to concede the importance of human *will* in economic processes, this was a determinate volition, not *free* will.[41] The distinction was important. All economic activity was underlain by a natural *will to live* that came into being the moment primitive organisms attained locomotion and cognition. But the form that this will to live took was determined by the environment in which the organism existed.[42] Kautsky illustrated this by drawing upon Engels' essay on 'The role of labour in the transition from ape to man' and demonstrating how the will to live took different forms for organisms at different evolutionary levels. For the lower organisms, whose living conditions were recurrent over generations, the purposive volition necessary for life became a habit, an instinct or a drive, passed on to their offspring, strengthened by natural selection and obeyed under all circumstances. For organisms whose living conditions were variable, instinct was not enough. A cognitive capability was required to apprehend and adapt to conditions. The process was partly Lamarckian, as the environment stimulated adaptation, and partly Darwinian, as natural selection eliminated the less intelligent.

For man, however, there was a crucial difference. As in Marx, the ability to make artificial organs – weapons and tools – external to the individual body, made man's relationship to nature unique within the animal kingdom. It also paved the way for a division that had no parallel in any other species – class society. Once extra-bodily organs existed, it became possible both to produce a surplus beyond everyday needs and for one group to own and control the means of production. This was not possible for any other species.[43] According to Kautsky, however, a parallel did exist. Forgetting Kropotkin and departing from Marx, Kautsky reverted to a Malthusian model and naturalized the class struggle. The struggle that ensued with surplus production was, he said, analogous to other intra-species struggles in which the animals contended for limited food resources. Both were characterized by antagonistic expressions of the will to live. Just as the strongest specimens of the species among predatory animals sought to bend the will of the weaker to their own needs, so the capitalists sought to satisfy their will to live by making the workers subservient. And the workers equally expressed their will to live through seeking to bend the will of the capitalists, either through demanding higher pay or ultimately seeking the expropriation of the expropriators. 'Class antagonisms', Kautsky concluded, 'are antagonisms of volition', expressions of the primal will to live.[44]

Conclusion

Thus the Marxist totems of revolution and class struggle were retained in Kautsky's thought only by forsaking the philosophical framework of Marxism in favour of a naturalistic account rooted in his understanding of Darwinism. The problem with this was that the linguistic and heuristic frameworks that Kautsky worked within were not neutral. By depending so heavily upon an evolutionary framework in which to conceptualize, and an organic language in which to express, socialism, Kautsky was inevitably led away from the Marxist heritage he sought to defend. As one critic of the use of scientific language in the politics of the left recognized, although both the Revolutionist and Reformist sought the Darwinian badge of honour, evolutionary language was not neutral: 'the organic analogy sanctions the Revisionist's procedure'.[45]

In fact, the organic analogy did more than provide a *post hoc* rationalization; it was integral to the construction and indispensable to the expression of such a politics. It was not the icing on the cake but the vital ingredient. Thus, by promoting this mode of reasoning, Kautsky eased the way for the very Revisionism he ostensibly opposed. Bernstein's nod to Darwinism, his evolutionary framework, his organic language, his absolute horror of utopianism had all been anticipated in Kautsky. The younger man was largely pushing at an open door, which Kautsky was both unable and unwilling to lock. His rebuttal of Bernstein merely drew a flimsy veil across the opening, the inadequacy of which was revealed when Kautsky came to pen his critique of the Bolshevik seizure of power.

With the actual example of Russia before him, Kautsky was forced to leave the realm of pure theory and explain his position in the light of events. His response, *The Dictatorship of the Proletariat* (1919), demonstrated how limited was Kautsky's notion of revolution and how close his position was to that of Bernstein. Today it is a rarely read reference point in the history of socialism, signalling the great divide between social democratic and communist interpretations of Marxism, and chiefly remembered for provoking Lenin's virulent polemic *The Proletarian Revolution and the Renegade Kautsky* (1919).

But Lenin's ire and feeling of betrayal were not warranted. Kautsky had never been the revolutionary that Lenin had imagined and the *Dictatorship* was not the work of a 'renegade' but the companion volume of the *Road to Power*. Bolshevism was accused of being Blanquist, utopian and transfixed by models of 1793 – exactly the complaints Bernstein had made against Marx and Engels.[46] Lenin was castigated as 'an empty phrasemonger', the worst crime in the Revisionist book.[47] And while Kautsky stopped short of following Bernstein into an outright rejection of the phrase 'the dictatorship of the proletariat', his treatment of democracy suggested that he shared Bernstein's view that this was the only route to socialism.[48] He still recognized, almost as an aside, that a 'revolution' would be necessary when the capitalists opposed democratic advance, but his conceptual framework was evolutionary and his language thoroughly organic.[49] Most tellingly, in a passage of which Bernstein would have wholly approved, the Bolsheviks were compared to a pregnant woman 'who performs the most foolish exercises in order to quicken the period of gestation'. 'The result of such proceedings is', Kautsky dryly concluded, 'as a rule, a child incapable of life.'[50]

8

Eugenics and Parasitology

Eugenics was the bastard son of Darwinism; or, to put it more correctly, the legitimate cousin. Darwin's cousin Francis (later Sir Francis) Galton coined the term – from the Greek root meaning 'noble in heredity' – in his 1883 book *Human Faculty*, but the essence of the eugenic argument had already been made in Darwin's *Descent of Man*.[1] It was here that Darwin noticed the paradox that civilized societies sowed the seeds of their own evolutionary destruction. As they advanced, so the 'weak' and the 'inferior' were exempted from the law of natural selection through poor laws, charities and medical treatment. This was a measure of evolutionary advance – to become more moral and compassionate was the noblest part of man's nature – but it also allowed those who would have been eliminated by the process of natural selection to breed unchecked. This threatened to send the evolutionary process into reverse.

As a cure, Darwin tentatively hinted at discouraging the poor and unfit from marriage.[2] His cousin was less restrained. If the 'unfit' were not being weeded out, and the 'fit' were not prospering, then the role of natural selection would have to be taken by artificial selection. Humans would have to become their own evolutionary selectors to save civilization from its innately destructive impulse. Society or the state, that is, would have to take on the role of those pigeon fanciers who populate the early pages of the *Origin*, selecting and accentuating certain traits and breeding out others from their stock.[3] Eugenics, therefore, took Darwin's diagnosis and added a prescription for the pathological condition of society said to arise from a confounding of the law of natural selection.

In a number of ways the eugenic analysis overlapped with a socialist exegesis of Darwinism – to such an extent that Galton's most important follower, Karl Pearson, defined himself as a socialist.[4] Firstly, eugenies was a collectivist doctrine. Whereas Darwin had left the unit of natural selection undecided between the individual and the group, the notion of racial fitness demanded that natural selection be a group process. Secondly, eugenics rested upon a positivistic faith in scientific rationality, and man's Promethean destiny to wrest control of his future, that echoed the agenda of most socialists. Thirdly, the instrument of eugenic policy, what Galton called the 'agencies under social control' intervening to improve 'the racial qualities of future generations either physically or mentally', was to be the state.[5] This chimed with the socialist objective of a strong state overcoming the interests of petty individualism. Fourthly, in the 'positive' versions of eugenics the state was to institute a social health programme that cared for mothers and babies.[6]

Thus although eugenics did not, as is sometimes asserted, resuscitate the Enlightenment dream of perfectibility, it did hint at a 'rational selection' that would dispose, once and for all, of *laissez-faire* individualism. It did so, moreover, by substituting the human mind, in the form of the state, for the vicissitudes of nature. As Galton put it, 'what Nature does blindly, slowly, and ruthlessly, man may do providently, quickly and kindly'. This, of course, was the essence of the socialist case. No wonder historians of both inter-war Britain and France have suggested that we reassess our attitudes towards eugenics, and learn to see it as a progressive force in politics.[7]

But it would be a mistake to assume that eugenics and socialism were ever truly compatible. At the heart of eugenics was a determination to close down one of the unresolved elements in Darwinism – that surrounding the relative importance of inheritance and environment – which had made its interaction with socialism possible. Eugenics rested on a hard hereditarianism that was irreconcilable with a belief in human freedom and equality. Darwin had been uncertain about the mechanism of inheritance and in the *Origin* had thrown his hands up in despair, sighing, 'our laws of inheritance are quite unknown'.[8] This provided an opening in which the socialist development of Darwinism insinuated itself.

Uncertainty about the relative significance of nature and nurture in determining an individual's character and abilities had left the door open to the Lamarckian and environmentalist explanations that we have seen recurring in the analysis of the socialists we have studied. Eugenists such as Karl Pearson slammed it firmly shut. The influence of Galton had encouraged Darwin to back away from Lamarckism in those passages in the *Descent* in which he discussed the inheritance of intelligence. Over the next thirty years the works of Weismann, and the rediscovered Mendel, combined with the *fin-de-siècle* fear of degeneration to gradually diminish the room for environmentalist explanations and enhance that of hereditarian ones. With typical mathematical precision, Pearson, the founder in 1904 of the Laboratory of National Eugenics, declared that a person's environment had less than one-sixth of the intensity of impact on their character and abilities than that of the hereditary influence of a single parent.[9] If this was so then an optimistic agenda of equality and environmentalism went by the wayside. This was what made the relationship between socialism and eugenics so problematic.

Our understanding of the complexities of this relationship has not been helped by the recurrent suggestion, most recently aired by Greg Claeys, that eugenics served the left as a means to revive utopianism.[10] Aside from sleight of hand being practised here, in seeking to tame an obviously awkward scientific doctrine by interpreting it in a more acceptable political framework, this seriously misunderstands the principal use of a eugenic argument for the left. It was not to provide hope for the future but to build a critique of the condition of existing society and to suggest that without reform the future would be worse. The most obvious impact of eugenics on left-wing literature was to spark a dystopian tradition, not to re-ignite a utopian one. Galton's wild imaginings of a future eugenic utopia, in unpublished novels such as *The Eugenic Community of Kantsaywhere*, did not set a template for the left.[11] The most popular literary products of the Fabian flirtation with eugenics were the bleak images of Wells' distinctly dystopian *The Time Machine* and *The Island of Doctor Moreau*. It was the fear of degeneration, if capitalism were allowed to continue unchecked – which we can also see in London's *Iron Heel* – not the hope of an evolutionary utopia that gave most succour to the left.

In any case, to confound the utopianism of eugenics with that of early-nineteenth-century socialism is disingenuous. Practically and philosophically these two variants had little in common. An Owenite utopia was a quick-fix solution by which a change of environment radically altered the traits and characters of the population. A eugenic utopia was to be created by breeding and inheritance, involving the sterilization and elimination of men and women deemed undesirable, and would be achieved only after a very long lapse of time. Moreover, to the extent that 'utopianism' was a pejorative term on the left, it referred to an ahistorical, mechanistic notion of change that was inimical to all evolutionary theories, eugenics included.

Our view has been further distorted by a number of studies of the Fabian embrace of the science.[12] These have tended to emphasize the sensational and sordid sexism and racism that often accompanied eugenics. But while the none too startling discovery that the Fabians were 'politically incorrect' may have played to the left-wing predilection for breast-beating and self-flagellation, it has told us little about the understanding of eugenics among the ordinary members of the mainstream left. To conclude from the example of the Fabians that the left, as much as the right, fell under the beguiling spell of eugenics is a serious misunderstanding. It not only does the vast majority of our predecessors a great disservice, but also tends to let the right off the hook by implying that it was somehow acceptable to be racist and sexist in this period.

In fact, as we shall see, it was possible for the left to engage with ideas of degeneration without succumbing to the worst excesses in the way the Fabians did. There was no question of socialists rejecting eugenics outright. As we have stressed throughout, late-nineteenth- and early-twentieth-century socialism was formulated and expressed in a

Darwinian discursive space. This gave it many of the same points of reference and boundaries and much the same medico-biological vocabulary as eugenics. But if there was no wholesale rejection, neither was there a wholesale acceptance. The vast bulk of mainstream socialists avoided the pitfalls of eugenics and the pratfalls of the Fabians, even while maintaining an ostensibly eugenic argument, rooted in the notion of degeneration but eschewing hard hereditarianism.

Parasitology

We need a new understanding of the relationship between eugenics and the left, and the key lies in parasitology. Parasitology is a strand in eugenic theory that has, until recently, been undervalued by scholars. But it was parasitology, rather than eugenics *per se*, which provided much of the left with its biological critique of capitalism and a biosocial prescription for socialism. Identifying this strand helps us explain how the mainstream left were able to seize on the critique inherent in the theory of degeneration – that present society was corrupt and failing – and to integrate the most attractive elements of the eugenic prescription, without succumbing to rigidly deterministic theories of inheritance. This was possible because the parameters of parasitology were quite different from those of eugenics proper.

Parasitology was an eclectic amalgam of selected ideas from Darwin, Lamarck, Spencer and Galton, which countered the strict hereditarianism of eugenics with a more sophisticated understanding of the interaction between an organism and its environment. From Spencer, parasitology took the assumption of a social organism. From Darwin, two lessons were drawn from the *Origin*'s discussion of the puzzling presence of parasites in the natural world.[13] The first was that parasites defied Spencer's law of differentiation. Rather than moving from the homogenous to the heterogeneous, parasites became less, not more, complex over time. They were, that is, atavistic. The second lesson was Darwin's Lamarckian explanation for this process. It was, he argued, a classic case of 'use it or lose it'. Parasites acquired – and lost – their characteristics in response to their environment. Finally, from Galton parasitology derived the notion that society was threatened by degeneration. But whereas Galton saw this as a case of evolution gone awry through the wrong types breeding, parasitologists, as the name implies, explained degeneration in terms of the presence of parasites that preyed upon the social organism. The distinction was crucially important. Galton's explanation made degeneration a hereditary process, a breeding of the 'unfit'. Parasitology explained degeneration in terms of an environment that created parasites.

Applied to human society, this produced strictures almost as sickeningly unpleasant as anything that spewed from a eugenist's pen, especially as the parasitologists often castigated the same outsider social groups, albeit with the helpful addition of bankers, financiers and capitalists. However, there was one crucial difference – parasitology did not write anyone off as hereditarily irredeemable. The condemnations of the 'underclass' that we find in parasitology, that is, were rarely couched in the same despairing pessimism that we find in Fabian eugenics. The imperative that flowed from parasitology was not to sterilize or to kill but to improve the environment. As we will see below in the case of Olive Schreiner, in parasitism the condition of degeneration could be countenanced as a counsel of hope.

Historians would do well to consider the extent to which this was the case on the left more generally. The comment of the character Owen in Robert Tressell's *The Ragged Trousered Philanthropists* (1914) that 'When we get socialism, there won't be any people like us, we shall all be civilised' is not, as Young has claimed, an example of 'unbridled pessimism and unqualified contempt for working people'.[14] It is, rather, an assertion that a changed environment will enable working people to cast off the chains that have stunted their growth and achieve their full evolutionary potential. This was some way removed from the simplistic environmentalism of Owenite socialism. According to Annie Besant, socialists had betrayed 'too great an inclination to think that environment is everything, and to ignore the reaction of the organism on the environment'.[15] But even if the transformation was to take time, there was still the hope that all could be redeemed in a healthy environment. In essence, Besant's argument that 'the evil tree brings forth its evil fruits' was as positive and environmentalist as Owen.[16]

The essential shared feature of eugenics and parasitology was not despair at the future of the degenerate, but a critique of existing society as dysgenic – that is, creating the conditions of evolutionary regress. In this both drew upon the argument of Edwin Ray Lankester's *Degeneration: A chapter in Darwinism* (1880). Lankester had made clear that evolution was not necessarily 'in accordance with a tacit assumption of universal progress'. Rather, the evolutionary process was always in one of three states: 'BALANCE, or ELABORATION, or DEGENERATION'.[17] Strictly speaking, however, the condition that parasitology diagnosed was one of deterioration rather than degeneration. The terms refer to distinct conditions of evolutionary regression but are easily confused.

'Degeneration' was first coined by the French physician Benedict Augustine Morel in 1857, to denote the spread of atavistic character traits through inheritance, and was intended to signify a morbid variation of inherited traits across generations.[18] Deterioration, by contrast, was a response to the environment, rather than to breeding. It referred to an evolutionary regression rooted in the external physical environment, rather than inherent character traits. But two factors allowed parasitologists to claim the degenerationist critique as their own. The first was a general blurring of the lines of demarcation between biological, social and moral arguments that characterized the late nineteenth century. This ensured that no one was too certain about delineating degeneration from deterioration, or either from decadence, and all three were freely confounded. The second was that degeneration was an even more ill-defined idea than Darwinism. There was no one defining 'degenerationist' text or theory and even the most die-hard eugenists, such as Pearson, had to admit that environmental factors shared an influence with hereditary ones. Thus it was possible to develop an ostensibly degenerationist account that actually gave pride of place to an environmentalist account rooted in the notion of deterioration.

The advantages of parasitology went beyond enabling socialists to exploit the degenerationist critique. It also appealed to an older sentiment on the left, which dignified labour and condemned idleness, and provided an emotive image with which to taint one's enemies. Both elements were present in Gronlund's condemnation of the 'parasites' preying on industry in his *Cooperative Commonwealth*.[19] The Greek root of the term 'parasitos', meaning 'one who eats at the table of another', might suggest that parasitism was no more than the biological rendering of the radical distinction between the 'idle' and the 'industrious' and compatible with Ruskin's moral elevation of labour. Certainly there was an element of this. Gronlund made full use of the imagery of parasites as repulsive creatures burrowing into the flesh, sucking blood and leaching upon the nutrients of the social organism. In case anyone missed the point, he even referred to the 'cannibalism' of the parasites and classed their financier friends as 'vampires'.[20] But for all this hyperbole, Gronlund's organicism led him to argue seriously that parasites were present in the 'partially evolved organism in which the arms and legs, and to a great extent the brain, are stinted in blood as much as possible'.[21] They were, he continued, a consequence of *laissez-faire* and a denial of the 'law of *Solidarity*'.[22] Just as surely as the untreated child's hair will contain lice, so a society that did not actively care for the social organism found itself the victim of parasites. Although Gronlund did not fully develop this argument, other socialists did.

Vandervelde and the social parasite

The most outstanding example of a socialist parasitologist was Emile Vandervelde (1866–1938). The writings in the 1890s of this Belgian future president of the Second International provided socialists with a detailed and scientifically credible means to engage with the degenerationist agenda – without succumbing to the fatalistic hereditarianism of eugenics. From his PhD thesis of 1891, which prophesied the coming of socialism in vivid metaphors drawn from the natural sciences,

Vandervelde worked at the rock face of a new cross-disciplinary research.[23] His aim was to replace, in 'modern and precise detail', Spencer's misty phrases about 'social organisms' with 'a concrete instance of that practical union and unison of Biology and Sociology, . . . so long proclaimed by the philosophers'.[24] Accordingly his writings were characterized by a free transference and cross-referencing of terms, categories and structures from the biological to the social and back again. And this pointed Vandervelde towards the problem of evolutionary regression.

He knew too much of contemporary biology to expect any evolutionary process – even that towards socialism – to be a simple, unilinear progression. In *L'Evolution régressive en biologie et sociologie* (1897), co-authored with Jean Demoor and the botantist Jean Massart, Vandervelde compared Belgian economic development to the struggle for survival and patterns of selection in animal and plant species. The pattern he identified was one of general progress and creation, threatened by the periodic dangers of regress and destruction.[25] By exploring the developmental parallels between social institutions and natural phenomena, in terms of an evolutionary theory of random adaptation and natural selection, Vandervelde highlighted the possibility of the degeneration of the social organism. This presented an opening for the social critic – a chance to argue that only socialism could save society. But it also laid a hereditarian trap.

The Lamarckian escape route Vandervelde took was to be found in his *Parasitism: Organic and social* (1895). This remarkable book made a simple non-eugenic argument in naturalistic terms. Parasites were a product of their environment and of habit. Thus to the extent that parasites were responsible for regress, a change of social environment, rather than an eliminative eugenics, would be adequate to reverse the trend. The parasite, both biologically and socially, was defined as '*a being which lives at the expense of another without destroying it and without doing it service*'.[26] This condition differed from both mutualism – where there was some exchange of services, however unfair – and predation, in which the predator sought to destroy its prey. The parasite, by contrast, depended upon the survival of the host but was unwilling and unable to exchange with it.[27] This made the relationship inherently unstable. Whether in the natural world or within human economic institutions, the parasite evolved to become ever more dependent and, as a consequence, to demand more of the host. This was an inevitable side effect of the atrophy of the organs that followed from disuse. At the same time, the host was weakened and enfeebled by the increasing demands of the parasite. The evolutionary imperative was that '[t]he society which is exploited by parasites becomes feeble; the parasitic individuals tend to be degenerate'.[28] The same process could be observed biologically or sociologically, with one significant difference: social parasites were of the same species as those they preyed upon.

There was no parallel for this in the organic world. And this made it easier to argue that social parasitism was the product of environment not inheritance. The three corollaries that followed from this one distinction between social and organic parasites all pointed to the lack of any inherited difference between the social parasite and its host. Firstly, whereas the organic parasite fastened on to an individual organism, the social parasite drew life from the whole community, the social organism.[29] Secondly, the social parasite, unlike the organic, did not consume its host – which would constitute cannibalism – merely its means to substance. Thirdly, although in organic nature the rigours of natural selection and inheritance played some role in honing the characteristics of the organic parasite, this was not the case for his social cousin.

The only possible conclusion to be drawn from the outward identity of parasite and host was that social parasitism was not an inheritable condition. It was an acquired characteristic, honed by habit.[30] A 'certain predisposition or aptitude for parasitic life may exist by organic inheritance' but in society the environment was 'the prime cause of parasitism'.[31] This was a reassuring message. It meant that the evolution towards parasitism was not inevitable or irreversible. If society were properly organised, it was 'possible for the social parasite to return to mutualist relations with his fellows'.[32] Indeed, this was not only possible, but also necessary. Where parasitism was allowed to continue unchecked, 'the collapse and total ruin of that society soon follows'. But with mutualism 'there will be a speedy elimination of the individuals and classes who become parasitic'.[33] Thus from a degenerationist launch pad, Vandervelde had side stepped the hereditarian snare and demonstrated the need for socialism as a means to save evolutionary progress.

What parasitism failed to provide Vandervelde with was a convincing basis for his ostensible Marxism. Vandervelde's science had preceded his socialism – it was his 'studies of Darwin's evolutionary theory [that] first led him to Marx's dialectic' – and this put his thought on a very different foundation to that of Marx.[34] Although never a Revisionist and unenthusiastic for Bernstein, like Bernstein, Vandervelde's socialism suffered from too little Hegel and too much Darwin. He stressed 'the organic character of the conception which serves as a basis for collectivist theories' and accordingly admonished Marx for the 'traces' of 'catastrophic utopias' that marred the *Communist Manifesto*.[35] Taken 'in its entirety', he argued, Marxism was an evolutionary ideology that understood socialism 'as the last term of the very evolution of capitalism'.[36] But as in Bernstein this rather disingenuously negated Marxism's revolutionary imperative. So did Vandervelde's tacit admission that Marx had been wrong about immiseration and his acknowledgement of Bernstein's role in 'dispelling the rather naïve illusion of certain socialists with regard to the rapidity of industrial concentration and the extent to which it has advanced'.[37]

Most worryingly of all, Vandervelde's parasitism eliminated the conceptual space for an inherent class struggle. The parasite may have been a powerful image with which to damn the capitalist – or any other group who opposed the advance of collectivism – but it had little in common with Marx's analysis of class antagonism. Parasitism was a problem for the sphere of exchange, not of production. In consequence, Vandervelde had remarkably little to say about capitalists or capital *per se* – as opposed to financiers and swindlers – and equated all groups who stood outside the social organism, regardless of whether they stood at the top or the bottom of society. Thus prostitutes were absurdly placed alongside bankers as parasites, with no consideration of their respective positions of power or differing relationship to capital.[38] This was disabling for any socialist analysis and bears comparison with the 'New Liberal' reworking of the concept of citizenship as a system of reciprocal rights and duties.[39]

Olive Schreiner and the parasitic case for female emancipation

The paradox of hope inherent in the condemnation of certain groups as parasitic on the social organism is nowhere better illustrated than in the writings of the South African author Olive Schreiner (1855–1920).[40] In *Woman and Labour* (1911) Schreiner, who spent the 1880s in London mixing in the circles of Besant and Aveling, made parasitology integral to her case for gender equality. In the process she condemned her own sex as parasites upon the social organism. In some powerful, but deeply disturbing passages, Schreiner deployed astonishingly unsisterly language to describe women as prostitutes and pustules of infection on the surface of civilization and as a deadly poison to the intellectual and moral environment of their race. Women, she argued, were physiologically deficient and morally bankrupt, intellectually inferior and prostituted to male sexual desire. Not only this but their very weakness threatened to pollute the blood of their male offspring, rendering them dangerously effeminate.[41]

The logic of this, Schreiner argued, was not eugenic condemnation. Rather it highlighted the urgency of female emancipation. Women were victims of their environment and their subjugation was a symptom of a deeper malaise that threatened the whole of western civilization. Like MacDonald in his essay 'The influence of sex on social progress', Schreiner recast the 'woman question' in evolutionary terms. But whereas MacDonald had made a Spencerian, progressionist case, Schreiner built her argument on the fear of degeneration. By using parasite theory to elucidate women's organic inferiority, she was able to demonstrate that the evolutionary development of women into parasites had become a barrier to future racial development. From this premise she proceeded to demand female emancipation in the name of evolution and eugenics.

For this feminist historians have judged her rather harshly. Sally Ledger, for example, has accused Schreiner of a dishonest 'exploitation' of the 'evolutionary metaphor' as part of a 'reverse discourse' rhetorical strategy of adopting the language of the oppressor in the interests of the oppressed. This procedure, Ledger concludes, was doomed to failure because it involved radically attenuating Schreiner's feminist vision without offering any immediate solution to women's subjugated status.[42] The crux of the complaint is familiar: Schreiner was clouding a clear political stance with a muddle-headed, instrumentalist use of science. But with Schreiner, as with Besant, Aveling, MacDonald, Kropotkin and a host of others we have studied, this is unfair and patronizing. 'Schreiner's parasitology was', as a recent study by Sarah Tasker has it, 'more than a cynical use of science in the service of ideology.'[43] Schreiner was genuinely and deeply indebted to Darwin, Galton and Spencer, and without them there was no purely political or philosophical argument, hermetically sealed from the corrupting influence of science, upon which she was able to call. Her approach to the question of gender relations was every bit as naturalistic as that of Darwin.

Schreiner's ideal of gender relations was inspired by the behaviour of a family of birds she had observed as a child in the African bush. The 'cock-o-veets', Schreiner declared, had reached a 'point of development' – which no human race had yet reached – of mutuality between the sexes, in both social and sexual relations. This went far beyond the 'noble function of reproduction' and expressed itself in 'distinct aesthetic, intellectual and spiritual functions'.[44]

Her account of the existing inferiority of women was also reminiscent of Darwin. In the *Descent* he had cited women's progressive removal from the theatre of struggle as an explanation of their evolutionary inferiority: men were still subject to the rigours of natural selection in the workplace and the free market but women, closeted at home, were rendered correspondingly soft.[45] Rather than dispute this, Schreiner accepted that woman's removal from the milieu necessary for mental and physiological development led to her evolutionary arrest. But whereas Darwin, Spencer and Galton were happy just to let matters rest here, with female inferiority proved, Schreiner took this as her cue for asking awkward questions. If the very process of civilization was to take half the human race out of the realm of natural selection, what were the eugenic consequences? When Darwin and Galton identified the same process among the 'inferior types' they fretted over dysgenesis. But surely, Schreiner ventured, the problem was far worse in the case of women? Thus, by applying the same eugenic criteria of 'fitness' to women as the eugenists applied to men, Schreiner was able to diagnose a pathological condition in society. The whole female sex had become parasitic and the irony was that this would ultimately destroy the very civilization that gave rise to this condition.

Schreiner was quite serious about this diagnosis. Despite – or perhaps because of – her own dependence upon a stipend from her brother, Schreiner was scathing about the consequences of the economic dependency of her own sex.[46] 'Normal' virtuous women, she declared, had been imperceptibly and anonymously reduced to the passive social position of sexual 'play things', whose only purpose was to titillate the appetites of the men upon whom they relied for existence.[47] The cause of this was passivity. The production of 'superfluous wealth' in society enabled certain groups to be removed from active production. In slave societies it was the slave-owners who were freed of the need to labour. In modern civilization it was women. In both cases it produced a 'debilitating effect'. Without the need to work, those placed in a condition of enforced passivity lost the means to do so. Instead they learnt to live off others. On a straightforward Lamarckian 'use it or lose it' model – of the kind Darwin had deployed when studying parasites in the natural world – those groups who were removed from the labour process could be expected to perfect the art of leaching from the workers. This is precisely what had happened to the female sex; they had evolved into parasites.[48] To fail to emancipate women, and reintegrate them into the labour force, would be to let the parasites continue unchecked until the point at which they destroyed their host.

Thus in one fell – or perhaps brilliant – swoop, Schreiner had transformed the 'woman question' into a biological mission to save civilization. To fail to grant women equality would have, and already was having, catastrophic consequences for the future of the race. This was quite a coup. Woman, who had played no more than an unproblematic reproductive role in Darwin, Spencer and Galton, was suddenly transported to top billing on the list of eugenic concerns. On Schreiner's account, just as 'a woman's cervix limit[ed] the size at birth of the child's head, so exactly, the intellectual capacity, the physical vigour, the emotional depth of woman, form[ed] also an untranscendable circle, circumscribing with each generation the limits of the expansion of the human race'.[49]

This was something more than a clever rhetorical twist to a eugenic argument. For a start Schreiner's understanding of the enervating consequences of passivity was as moral as it was physical. For Schreiner, labour was not simply a matter of physiological activity for survival. It was also a working out of a healthy psychology – what Schreiner called 'will' – and part of the natural, organic expression of what it was to be human.[50] Moreover, Schreiner's test of the level of a society's 'fitness' was far more nuanced than the simple measures of military or economic power found among male degenerationists. Even among the 'cock-o-veets' Schreiner had looked for less quantifiable factors, including the emotional, spiritual and intellectual quality of life. But the key element that marked off Schreiner's parasitology from what she called Galton's 'ludicrous' eugenics was the Lamarckian model of evolutionary transmission on which her theory rested.[51] To the extent that inheritance, rather than environment, played any part in her critique, it was only in the form of a 'soft' theory of a progressive disuse of their faculties, inherited as an acquired characteristic. This was quite some way from the hard hereditarian theories peddled by eugenists.

9

The ILP and the Socialist Library

There is nowhere better to study the dissemination of a Darwinian-influenced version of socialism, and the slow turning away from it in the 1920s, than in the political education activities of the Independent Labour Party (ILP). The party represented the mainstream of British socialism and its largely self-educated membership, lacking either the rigid dogmatism of their rivals in the Social Democratic Federation or the urbane insouciance of the Fabian Society, greedily sought a scientific sanction for their politics. The ILP were not the straightforward purveyors of ethical socialism they are often portrayed as. From the party's defining statement, *The ILP: What it is, and where it stands* (n.d.), the ILP's mind-set was evolutionist and its language organic.[1] Indeed, when J. Bruce Glasier came to explain the need for the ILP to his membership he did so in distinctly Darwinian terms: 'Why, we may ask ourselves, does the ILP exist at all? Why did it come into being? Because there was a need for it. What is the reason it has grown so remarkably? Because of its vitality – because of its fitness to survive.'[2]

But what the ILP did retain from its roots in radical non-conformity was a democratic epistemology which demanded that if Darwinism was to be the basis of the socialist argument then this must be understood by all of its members. There was no question of following Jack London, with his cadre of Nietzschean 'blond beasts' leading the way to socialism, or Kautsky, who wanted the bourgeoisie to provide the proletariat with scientific leadership. In the ILP all socialists had to be given access to scientific knowledge. And this made the party's political education work, with MacDonald at the helm, of primary importance. MacDonald wanted to make socialists in his own image. Not 'roysterers' or 'brawlers', the 'windbag species' of socialists, but 'hardworking, steady men . . . who, feeling their cause to be good, nourish and nurture its success with the same patient care as they nourish and nurture their children'.[3] These 'scientific socialists' had to be taught to think in an evolutionary framework and express themselves in an organic language.

This political education work has generally been underestimated by historians who have seen the ILP as an 'electoralist' organization, trapped in a specifically British 'ethical' socialist tradition and indifferent to the task of 'making' socialists. In this their conclusions have been shaped by their research agendas as much as by the available evidence. Ever since E. P. Thompson's 'Homage to Tom Maguire' it has been *de rigueur* to claim that the ILP is best studied at a local level.[4] Consequently, we have gained a tremendous amount of detail on the activities of local branches but concomitantly few studies of the ILP as a national organization and none with a detailed consideration of its pre-1914 ideology.[5] The necessarily narrow focus of local branch studies – and resultant neglect of the broader ideological position of the national party – has exaggerated the ILP's dependence on a purely 'ethical socialism'.[6] And with ethical socialism by definition quintessentially 'British', the circle has closed with the ILP an insular organization, cut off from more general trends in European thought.[7]

Our approach is different. In place of the insular, ethical and electoralist portrayal of the ILP, our argument about the relationship of Darwinism and socialism suggests a more outward-looking and propagandist party, committed to an evolutionary view of socialism. Throughout we have argued that the interaction of Darwinism and socialism profoundly affected the language and ideology of the British left. This interaction created a socialism that was distinct from the liberal and radical traditions on one side and Marxism on the other. Far from being cut off from the European mainstream, this placed ILP socialism within the same discursive space as framed the Revisionist controversy. Equally, it rendered it necessary for the ILP to undertake the work of 'making socialists'. Its propaganda efforts may have fallen some way short of the missionary activity envisaged in Robert Blatchford's *Clarion*, but there was a determined effort to develop an evolutionary mindset and instil an organic language into the rank and file.[8]

We are not arguing that an ethical element was absent from ILP socialism, but that its importance has been overestimated. Historians have been insensitive to the degree to which ethical socialism was increasingly recast in an evolutionary framework and expressed in an organic language, rather than in the rhetoric of the pulpit. This movement was not restricted to MacDonald. Even Keir Hardie, the living embodiment of ethical socialism, came to argue that socialism was 'no longer the hope of the visionary with his head high up in the clouds . . . [but] an organic evolution of a higher social order'.[9]

Typical of the ILP approach was the educational campaigner Margaret McMillan. Described by her most recent biographer as 'an ethical and evolutionary socialist', McMillan's politics were inspired by both a dissenting Christian ideal and Henry Drummond's *Lowell Lectures on the Ascent of Man* (1894).[10] Drummond's argument, that the Darwinian struggle for life had developed into a struggle for the love of others, provided McMillan with an account of altruism which encouraged her view that socialism was an inevitable evolutionary outcome. In addition, McMillan's friend and neighbour Kropotkin provided her and many other ILP members with a model of mutual aid that encouraged them to see evolution and ethics as compatible.[11] Even the overtly religious Philip Snowden, who outlined his socialist vision in *The Christ that is to be* (1903), attended lectures by Kropotkin in 1889 and regarded socialism as an evolutionary advance from capitalism.[12] This almost insidious building of evolutionary concepts and an organic language into the socialist argument was far more effective in shaping the thinking of the ordinary membership than any explicit discussions of the relationship between socialism and Darwinism.

The only ILP pamphlet to address this question specifically was, in any case, a rather confused affair. Laurence Small's *Darwinism and Socialism* (1908) began rather bizarrely with a bibliography pointing the reader towards Darwin's *Descent*, Ferri's *Socialism and Positive Science*, Wallace's *Darwinism* and Kropotkin's *Mutual Aid*. Although Spencer was not included in the recommended reading, the definition of evolution with which the text opened – 'an unfolding from the less to the more perfect, from the simple to the complex' – was no more than a restatement of Spencer's law of differentiation.[13] But, unlike in Spencer, the prospect of regress was prominent in Small's definition. Evolution, he said, could be positive, negative or neutral; it was no more bound to go forward than backward, individually or socially. This was why socialism was necessary.

The example of H. G. Wells' *When the Sleeper Wakes* (1899), in which the hero awakes 200 years in the future to find the world an immense jungle, was invoked by Small as a warning 'of what evolution and laisser faire may do for us'.[14] The message – that the danger of degeneration lurked in individualism and that socialism was its antidote – was clear enough, but Small got confused when he tried to answer Haeckel's accusation that socialism would entail an abrogation of Darwinism. He floundered between two different arguments. The first was the naturalistic case that individualism upset the operation of true natural selection and that socialism would restore its operation. To justify this he drew upon W. K. Clifford, who had argued that any society that encouraged the development of the self against the common interest would disintegrate, and Darwin's argument that the moral sense has itself developed through natural selection. On this point, Small concluded that 'the social instincts are of a more permanent and of a more deeply seated character than the self regarding' and recommended chapters III, IV and V of the *Descent* to anyone who thought Darwinism anti-socialistic.[15]

His second argument was that the law of natural selection could be circumvented, and its operation rendered less relevant, in the cultural spaces that man occupied. This was a non-naturalistic argument that stressed man's superiority over the 'lower animals'. While natural selection was no more open to alteration than the law of gravitation, it was possible for men to make natural laws 'subserve the purpose of man, and not entirely be his master'. Just as the building of a bridge defied the law of gravity, so it would be possible for man to defy the law of natural selection.[16]

The Socialist Library

A far more important element in the ILP's political education programme was the Socialist Library, a series of twelve books published between 1905 and 1919, and edited by MacDonald. The Library was the only series of full-length books that the ILP published before 1914 and, prior to the launch of the *Socialist Review* in 1908, was the primary ILP outlet for the discussion of socialist theory. It also provided the key forum for introducing lay activists to the movement to develop Darwinian socialism. However, the Library has hitherto escaped the notice of historians, due to the tendency to dismiss the ILP as ideologically uninteresting.

According to its *Prospectus*, the Library had three aims. The first was to remedy what MacDonald called the 'deplorable lack' of socialist literature produced in Britain' and thus achieve 'a firmer hold [of socialism] upon the intellectual classes and amongst Socialists themselves'. Its second aim was to emulate the continental example where, it was noted, vigorous socialist movements were complemented and reinforced by an equally vigorous socialist literature. To this end, it pledged to publish studies of both socialist theory and specific social problems. Thirdly, the Library aimed to introduce British socialists to 'translations of the best works of foreign Socialists'; 'for it is to be regretted that Socialists in this country have been cut off so long from the foreign literature of the movement'.[17] Their reintegration, MacDonald hoped, would prove doubly salutary: breaking the British socialist's 'happy-go-lucky disregard for theory' and countering the presentation of continental theory, 'through reddened spectacles', as the 'cause of revolution'.[18]

The launch of the Library represented an important new stage in the ILP's propaganda. Founded in 1893, it was not until its Fourth Annual Conference in 1897 that the party established a Publications Committee – with MacDonald as a co-opted member – and adopted a policy of publishing in support of its aims. Even then it limited itself to a monthly report for members, which became *ILP News* in 1898, and to the production of a few small pamphlets and single-page leaflets. The take-off in the ILP's publishing activities came only in 1901, with the resolution of the National Administrative Council (NAC) to undertake a propaganda drive that would build upon the prevailing anti-Boer-War and pro-social reform feeling in the country.

These practical campaigns, around issues such as pensions,

Table 9.1 The Socialist Library, edited by J. Ramsay MacDonald

Volume I	Enrico Ferri, *Socialism and Positive Science (Darwin–Spencer–Marx)*, London: ILP, 1905. By Enrico Ferri, professor of penal law in the University of Rome. Director of the *Scuola Positiva*. Translated by Edith C. Harvey, from the French edition of 1896.
Volume II	J. Ramsay MacDonald, *Socialism and Society*, London: ILP, 1905.
Volume III	Jean Jaurès, *Studies in Socialism*, London: ILP, 1906. Translated with an introduction by Mildred Minturn.
Volume IV	Sydney Olivier, CMG, *White Capital and Coloured Labour*, London: ILP, 1906.
Volume V	Emile Vandervelde, *Collectivism and Industrial Evolution*, London: ILP, 1907.
Volume VI	Philip Snowden, *Socialism and the Drink Question*, London: ILP, 1908.
Volume VII	Edward Bernstein, *Evolutionary Socialism: A criticism and affirmation (Die Voraussetzungen des Sozialismus und die Aufgaben der Sozialdemokratie)*, London: ILP, 1909. Translated by Edith C. Harvey.
Volume VIII	J. Ramsay MacDonald, *Socialism and Government*, 2 vols, London: ILP, 1909.
Volume IX	Margaret McMillan, *The Child and the State*, Manchester: National Labour Press, 1911.
The ILP Library*	K. Kautsky, *The Dictatorship of the Proletariat*, Manchester: National Labour Press, 1919. Translated by H. J. Stenning.
	F. H. Hayward, D.Lit, B.Sc., and B. N. Langdon-Davies, MA, *Democracy and the Press*, Manchester: National Labour Press, 1919.
Volume XII	J. Ramsay MacDonald, *Parliament and Revolution*, Manchester: National Labour Press, 1919.
Extra Volume	*The Revolution in the Baltic Provinces of Russia: A brief account of the activity of the Lettish Social Democratic Workers Party*, by 'An Active Member', London: ILP, 1907. Revised and prepared for publication by Ernest O. F. Ames.

* See the explanation on p. 104.

unemployment, housing and temperance, were successful enough to prompt the launch of a weekly propaganda leaflet, entitled *The Platform*.[19] By 1904, following the purchase of a majority shareholding in the *Labour Leader* newspaper the previous year, the Publications Department was showing its first profits and the NAC was claiming that '[e]xperience has shown that it should be possible to dispose of a first edition of at least 50,000 copies of every pamphlet we publish'.[20] Indeed, the print run for Philip Snowden's four-page pamphlet, *The Aims and Policy of the ILP*, exceeded 100,000.[21] Flushed with this success the ILP launched the Library as its first full-length book series.

The Library honoured its undertaking to publish both specially commissioned works by domestic socialists and translations of key continental texts, dividing its output almost equally between the two. It similarly kept its pledge to publish studies of both socialist theory and specific social problems. Ferri's *Socialism and Positive Science* (1905), MacDonald's *Socialism and Society* (1905) and Bernstein's *Evolutionary Socialism* (1908) were primarily theoretical texts, whilst McMillan's *The Child and the State* (1910) and Snowden's *Socialism and the Drink Question* (1908) concentrated on specific social problems. Other volumes, particularly those by Jaurès, Vandervelde and Olivier, and MacDonald's *Socialism and Government* (1909), offered a mixture of ideology and policy.

It was a mixture that proved popular. If sales are any indication then the hope of the NAC, 'that as many . . . members as possible – especially the younger members and the speakers – . . . study these volumes, so that they may gain a firm grip on the principles and work of the Socialist movement at home and abroad', was fulfilled.[22] Ferri's *Socialism and Positive Science* was into its fifth edition by 1909 and Vandervelde's *Collectivism and Industrial Evolution* (1907) went through five editions in just two years. Most successful of all, however, was MacDonald's *Socialism and Society*. It reached its sixth edition as early as 1908 and was still being published in 1919. It was even translated into a number of European languages, including German and Norwegian.[23] What united all the contributors was the evolutionary framework in which they presented their socialism and the organic language they deployed. MacDonald chose the texts carefully to reinforce what he saw as the most essential characteristic of ILP propaganda: 'We believe in evolution. We have made "the Social Revolution" a hackneyed shibboleth on our platforms, and it has had no more influence in our modes of action.'[24]

The 1907 ILP Conference reported 'a considerable market for Socialist and Socialistic literature', but although the early volumes of the Library were in almost constant reprint, the initial momentum could not be maintained.[25] From 1908 the Library experienced a downturn and by 1911 it had ceased to trade. A number of projected volumes – including titles such as *Socialism and Religion*, *Socialism and Foreign Affairs*, *Progress of Socialism in England*, *Socialism in Parliament*, *Socialism and the Rural Population* and *Socialism and the Position of Women* – never appeared. The first hint of trouble came with the failure of an ill-judged 'Extra Volume' to raise money for socialist comrades in the Baltics.[26] But the real difficulties began in 1910, with the ILP's costly mistake of attempting to emulate the 'Vorwarts' Printing Press in Berlin by establishing the National Labour Press in Manchester. The relocation from London disrupted the printing and delivery of publications, as well as breeding resentment within the party, at precisely the moment when sales of socialist literature were beginning to dip.[27] Consequently, a small loss in the trading of the Publications Department in 1909 became a gaping deficit by 1913, when the NAC reported that 'sales of literature are so small that the publication of any pamphlet is unsafe'.[28]

The Library was briefly revived in 1919 to publish the twelfth and final volume in the series, MacDonald's *Parliament and Revolution*. What had happened to volumes X and XI is unclear. Two books, Kautsky's *The Dictatorship of the Proletariat* and Hayward and Langdon-Davies' *Democracy and the Press*, had appeared in a new series entitled 'The ILP Library' in 1919 and although this was not a renaming of the Socialist Library, it does seem legitimate to regard them as the missing volumes.[29] Not only do the two works fit sequentially, but more importantly they are compatible with the very distinctive identity that the Library had developed.[30]

Socialism and Positive Science

The title of volume I in the Library, Enrico Ferri's *Socialism and Positive Science (Darwin–Spencer–Marx)*, captured perfectly the spirit of the entire series. MacDonald's description of Ferri's aim – 'to show that Darwinism is not only not in opposition to Socialism, but is its scientific foundation' – could have stood as the Library's motto. Ferri was a professor of penal law in Rome, a radical deputy between 1886 and 1901 and a sometime collaborator of criminal anthropologist and theorist of degeneration Cesare Lombroso.[31] He was also, he told his readers, a 'convinced follower of Darwin and Spencer' but, finding they stopped 'short halfway', he had turned to Marxism for 'the practical and fruitful complement in social life of that modern scientific revolution'.[32]

Despite giving Darwin and Marx equal billing, it was Ferri's development of Spencer that provided the basis for the book. This was no more profound than that of Gronlund or Besant, but his international appeal – *Socialism and Positive Science* was translated into French, German and Spanish the year after publication, and appeared in the USA in 1901 – made Ferri an important figure in the development of a Darwinian socialism. The book opened with Ferri recalling the 1877 spat between Haeckel and Virchow, and took its lead from Virchow's suggestion that 'Darwinism leads directly to socialism'.[33] To prove this he needed to refute Haeckel's threefold case that Darwinism was fatal to the socialists' hopes. In the process, Ferri redefined socialism in an organic language and self-consciously defined his 'scientific socialism' in opposition to the 'sentimental socialism' of the pre-Darwinian period.

Thus on Haeckel's first point, that while socialism strove for equality, Darwin naturalized inequality, Ferri disputed the charge. Scientific socialism, he said, 'inspired by Marx', had 'never denied the inequality of individuals as of all living beings'. If socialists, such as Bebel in *Woman in the Past, Present and Future*, had done so, that was an unfortunate 'mental habit' of sentimental socialism. The newer, scientific version did not deny an inequality of 'innate and acquired abilities' but sought only to remove those social inequalities that accentuated them.[34] On Haeckel's second point, that socialism wanted all to succeed while natural selection doomed the majority to perish, Ferri made a more robust defence. He agreed that the 'struggle for existence' was an 'iron law' and mocked those socialists who thought that this law could 'fall inert at the feet of man as if he were not an indissoluble link in the great biological chain!'. But, drawing on Spencer and Kropotkin, Ferri maintained that once food was secure, the law of solidarity assumed as much importance as that of struggle, and that, in any case, the struggle for life took an increasingly peaceful and intellectual form.[35] Finally, to the assertion that natural selection was necessarily aristocratic in operation, Ferri retorted by accusing Haeckel of a teleology that confused the survival of the fittest with the 'survival of the best adapted' and argued, like Wallace, that only socialism could ensure genuine natural selection.[36]

Ferri's debt to Spencer can be seen in all aspects of his thought. The organic analogy, the law of differentiation, the notion of progress and the idea that the form of natural selection changes over time were all learnt directly from the English philosopher. And despite Ferri learning the limitations of utopianism from Marx, it was Spencer's organic analogy that underlay Ferri's critique of the sentimental socialists. Their failing, he said, citing Spencer, was to imagine 'that human society is like dough, to which law can give one form rather than another without taking into account the qualities, tendencies and aptitudes, organic and physical, ethnological and historical, of different peoples'.[37]

Two significant differences between Ferri and Spencer, however, are worth noting. Firstly, Ferri's notion of progress was less straightforward than that of Spencer. Whereas Spencer only reluctantly came to admit that evolution could move backwards as well as forwards, this is a far more prominent theme in Ferri – as we might expect from someone who had worked with Lombroso. Instead of a unilinear progression, Ferri wrote of 'the eternal rhythm of living humanity': 'there has really been progress, amelioration on the whole, not following a straight ascending line, however, as Goethe has said, a spiral with rhythms of advance and retrogression, of evolution and dissolution'. This was not a closed circle, but 'a spiral which seems to come back on itself but which always advances and rises'.[38]

The second significant difference was in Ferri's attitude to individualism. As with London, Gronlund, Besant and MacDonald, Ferri read Spencer's organicism as an implicit critique of individualism. Biology had shown that the individual was an aggregate of simpler elements, just as sociology confirmed that the individual could exist only as a member of society. This convinced Ferri that 'at the end of the nineteenth century the individual, as a being in himself, is dethroned in biology as in sociology'.[39] And this interdependence, taken with the law of differentiation, allowed the scientific socialist to 'affirm, with mathematical certainty' that the trajectory of evolution was towards an ever more socialistic state.[40]

The role of Marx in Ferri's thought was more complex. Ferri was unreserved in his praise for Marx, whom he credited with applying 'the theory of evolution in the economic domain' and whose work, he said, completed that of Darwin and Spencer.[41] But when it came to details Ferri was a lot less certain. His defence of the notion of revolution, for example, was no more convincing than that of Kautsky. Socialism would come about 'possibly after some excess of fever – from the noxious products of the present phase of civilisation', but this organic sickness was more likely to be the stamping of the iron heel than a creative action.[42] Revolution was, in any case, only ever possible as the culmination of a long period of evolution. 'Human society, forming a natural and living organism, like other organisms, cannot undergo sudden transformations as those imagine who think we must resort only, or by preference, to revolt and personal violence to realise a new social organisation.'[43] To hope otherwise was as foolish as a young man attempting to pass through puberty in a single day. Moreover, Ferri suggested that the likelihood of revolution declined in line with anthropological advance. The 'old method of revolutionary romanticism', he suggested, had been all but eliminated in Germany and the Anglo-Saxon nations and was now only suitable for the Latin nations, at a lower stage of evolutionary development.[44]

An anthropological argument also influenced Ferri's attitude to Marx's economic determinism. Ferri approved of understanding economic phenomena as a basis of all other human and social manifestations, because this seemed to correspond with the biological law that function was determined by organ. But he argued it was important not to be dogmatic in asserting that economics was the exclusive foundation of social mores and practices. Marx's economic determinism, that is, needed to be balanced against the physical determinism of the natural environment and the anthropological determinism of racial categories.[45] He thus suggested, to a far greater extent than Marx, that there were naturalistic limitations on the free development of human history.

Eugenics and degeneration

Ferri's concern to balance Marx's economic determinism with an anthropological determinism rooted in racial categories, and his recognition that degeneration was a potential in organic development, were discordant notes in the Darwinism that the Socialist Library purveyed. On the whole the Library's volumes are remarkably free of the racist politics, degenerationist assumptions and eugenic prescriptions that we have come to associate with the Fabian embrace of Darwinism.

There were, of course, exceptions – most notably volume IX, MacDonald's *Socialism and Government* – about which more in a moment. And it would be wrong to suggest that theories of degeneration and eugenics failed to permeate the ILP at all. Small's pamphlet *Socialism and Darwinism* was, as we have already noted, premised upon the notion that socialism was necessary to save society from degeneration, and towards the end even hinted at a eugenic prescription. Artificial selection, said Small, could be applied 'to human beings, even as it has been applied in horticulture and agriculture, and in the breeding of animals'.[46] One could discourage or forcibly prevent the marriage of the congenitally diseased and deformed, in order to prevent the breeding of 'those reversions to anti-social type forming the genuine criminal ingredient of humanity'.[47] Nor was this the only occasion on which the ILP addressed the question of eugenics. Under MacDonald's editorship, the *Socialist Review* regularly printed pro-eugenic articles by Sidney Herbert and Maurice Eden Paul.[48]

But it is important, as we said in the last chapter, to keep this flirtation in perspective. The *Socialist Review* just as regularly published anti-eugenic articles by F. C. Constable and Lancelot T. Hogben,[49] and MacDonald occasionally wrote editorials criticizing eugenists and rejecting Lombrosian criminal anthropology.[50] The balance of articles and editorials was, as Herbert complained, firmly in favour of the 'average socialist' who accepted environmental and economic explanations over hereditarian ones.[51] And on balance this was also the position of the ILP in general and the Socialist Library in particular.

The three British socialists (excluding MacDonald) who wrote for the Library, McMillan, Snowden and Sydney Olivier, each demonstrated a consistent commitment to the traditional radical and socialist acceptance of the power of nurture – of education, environment and equality – over the power of nature: innate ability, inheritance and inequality. McMillan's *The Child and the State* was an extended protest against notions of inherited aptitudes and a plea for an acceptance of both the equality of all children and the need for adult education. In this she was at pains to distinguish her educational programme from that of the more eugenically minded National Efficiency advocates, who would provide only a 'technical education' to make 'drudges' for the ruling class.[52]

Snowden's *Socialism and the Drink Question* was equally clear in embracing an environmentalist explanation of social problems. He treated 'the Drink Question [as] but one phase of the Social Problem', and found an answer in the municipalization of the drink trade. He was aware of medical opinion linking alcohol and degeneration, and admitted the possibility of a 'dormant hereditary disposition' to drink, but paid them little mind, preferring environmental explanations, rooted in social and economic conditions, to hereditarian and biological accounts of alcoholism.[53] Most strikingly of all, Olivier, as his title *White Capital and Coloured Labour* indicates, presented an economic analysis, in which the categories of 'capital' and 'labour' were treated as more important than those of 'race' and 'colour'. Thus McMillan asserted education, Snowden environment and Olivier equality, against the eugenic values of innate ability, inheritance and inequality.

Even MacDonald's wavering from this non-eugenic and non-degenerationist line betrayed confusion and indecision rather than offering a fully worked-out position. *Socialism and Government* was by far his most eugenically inclined work, but even here his argument veered between an endorsement of a moderate eugenics and the denial of its possibility. Throughout, bold pleas to make eugenics 'a matter of State concern', and apocalyptic warnings about the dangers of race deterioration, fizzle out into the relatively benign recommendation that the state reinvigorate sexual selection by directing personal taste in beauty, to ensure the propagation of 'healthy and comely men and women'.[54]

MacDonald was undeniably interested in theories of inheritance – recommending studies of Mendelian texts to readers of the *Socialist Review* – and briefly toyed with eugenics.[55] This went some way beyond general statements about seeing the signs of degeneration all around, which, as we have seen, were the common currency of those who wanted to argue that only socialism could save evolution.[56] But in his editorship of the *Socialist Review* MacDonald was noticeably even-handed on the issue of eugenics – though significantly not on other issues – and this may be indicative of the fact that he was at best only spasmodically convinced by his own eugenic leanings. Much like the socialists in Chapter 8, when it came to the crunch MacDonald's argument was Lamarckian and environmentalist, not eugenic and hereditarian. As he explained:

> Whilst the individualist and the reformer offer changed systems of Poor Law administration, segregation of the unfit, the lethal chamber, and similar things as preventives, the Socialist regards race deterioration as a social phenomenon, the result of general ill-health, an organic disease undermining the system.[57]

Similarly, although there are undoubtedly hints of a rather unpleasant racial analysis in MacDonald – he warned that racial antagonism was 'in the blood' and feared that cross-breeding among any but the most closely aligned races, such as the Danes, north Germans and Celts, was liable to create a weak and unhappy race – these were not recurrent themes.[58] Nor did MacDonald show much conviction in attempting to propagate them. On the contrary, for the Socialist Library he commissioned the 'avant garde and courageous' Olivier, rather than any author closer to the 'overwhelmingly racist consensus' of Eurocentric and Kiplingesque views found among many of his old Fabian comrades.[59]

Sydney Olivier

Sydney Olivier (1859–1943) had been one of the original Fabian essayists and a member of the Rainbow Circle in the 1890s. Born in the year the *Origin* was published, as a young man Olivier had quenched his thirst at the same intellectual fountain as the other socialists we have studied, greedily drinking down Darwin, Spencer and a host of other evolutionists.[60] Their influence can be seen in Olivier's major work, *White Capital and Coloured Labour* (1906). However, the horror of the 'non-adult races' and fear of degeneration through interbreeding which suffuse the writings of his fellow Fabians H. G. Wells and Sydney Webb are absent from Olivier's exploration of race 'from the point of view of evolutionary biology'.[61] Instead, Olivier argued strongly that racial discrimination was unjustified and that interbreeding held the key to a strong organic community. He reached these conclusions during his stint as Colonial Secretary for Jamaica, between 1899 and 1904. Defying Aveling's expectation that exposure to the non-white races was bound to convince someone of a racial hierarchy, Olivier's time on Jamaica disabused him of his belief that degeneration would arise from interracial breeding. Upon his return to England he took the opportunity to develop his ideas in a series of articles and in his well-received and 'widely read' book.[62]

White Capital and Coloured Labour was concerned with the interaction of races in the industrial system and, in particular, with the conflicts and misunderstandings that arose from the imposition of industrialism upon the populations of the West Indies and Africa. In part, Olivier was revisiting the debate between John Stuart Mill and Thomas Carlyle half a century previously, in which the character of the 'lazy' black, recalcitrant in adapting to life as a wage-slave, had been contested.[63] Olivier too was concerned to examine why the black African and West Indian seemed so ill-suited to the labour required of them by white capital. But he also drew upon more up-to-date influences. In particular, the hand of J. A. Hobson can be discerned in the centrality Olivier gave to the importance of the 'economic motive' for the spread of empire.[64] Despite his own Christian beliefs, Olivier had no time for the philanthropic cant of those religious evangelists who made a 'white's man burden' argument.[65]

The most interesting and distinctive feature of Olivier's argument, however, was that it was built 'on Darwinian principles'.[66] At times this threatened to drag Olivier into the kind of anthropological account of race that we have seen in Huxley and others. This danger is obvious in those passages in which Olivier asserts that 'the white man's civilisation is a higher and better thing than the black' and commends the spread of the industrial system as a beneficial influence upon the 'savage' African.[67] But what set Olivier apart from other Darwinists was that he avoided confounding the social conditions of 'civilisation' and 'savagery' with the racial categories of 'white' and 'black'. This allowed him to condemn the 'savagery' of Africans and praise the 'civilisation' of Europeans, without simultaneously identifying the virtues or failings of a social system with the biological make-up of their populations. There was, he argued, no 'racial significance in the condition of savagery' or 'civilisation'.[68] This may have been disingenuous, and Olivier has rightly been accused of paternalism, but it is striking that amidst the racism that seemed to trail in the wake of Darwinism, Olivier was able to make an evolutionary argument for racial equality.[69]

He achieved this by attributing all racial differences to a Lamarckian response to the demands of different climates and environments upon the human organism. In this way, he robbed the concept of race of its restrictive power of inheritance and was able to dismiss any differences as amenable to change through education and a changed environment. In place of 'the unalterable limitations of the racial faculty' cited by other Darwinists, Olivier found only acquired characteristics that would 'yield, more or less, to educational influences'. Racial differences, to the extent that they existed at all, were rendered inessential 'excrescences or shortcomings of Humanity'.[70] Instead of dwelling upon them it was more illuminating to look both at the 'sparkling variety' within racial categories and the essential identity of all humans, 'transcending Family, Race, and Nation alike'.[71] The latter included traits such as 'the musical ear' and the 'very specialised and elaborate human faculty and achievement in Art, Science, Philosophy', which were found across races but, as Wallace had noted, offered no advantage in terms of natural selection.[72] Racial differences, by contrast, were shallow, reversible and easily explained in Lamarckian terms. The power of inheritance in restricting the achievements of an individual of any race was set almost to nought. There was, Olivier concluded, no political or human distinction for which blacks were disqualified by African blood.[73]

Not that Olivier was uninterested in, or underestimated the importance of, inheritance. His solution to the problem of the future relationship between white capital and coloured labour rested on biological inheritance. If white and black were to exist side by side – 'fused into one organic community' – then two things were needed. Firstly, the 'virus' of 'negrophobia' that blighted South African society and the southern US states had to be rejected as 'an extremely destructive element within the social organism'.[74] Secondly, 'race fusion', by which Olivier meant 'the bodily process of blending by intermarriage, or by some alternative physical process of establishing sympathetic understanding', must occur.[75] The example of the sizeable 'hybrid class' in Jamaica, who had 'attained an organic and honourable position in a mixed community', convinced Olivier that hybrids were 'advantageous' to society. The hybrid, he argued, was to be celebrated as 'a superior human being', rather than feared as a source of degeneration.[76]

This implied an agenda that was both racially and sexually radical. The sexual radicalism was tempered somewhat by Olivier finding a 'good biological reason' for the 'social objection' to white women marrying black or coloured men.[77] But he held fast to his belief that the 'interbreeding and mixture of races' – through white males and black females – would produce more intelligent and physically stronger specimens. And the cross-breeding was to be an equal intermixture – not what we might crudely call a 'white brain in a black body' scenario. Olivier's hazy account of the 'physiological aspect' of interbreeding made the process analogous to candy-pulling, in which the strands became inextricably intertwined at every level and would lead to a doubling of the 'race element' in every cell.[78]

Thus despite such quirky views, and the odd discordant note of confusion, Olivier made an evolutionary argument for racial equality. Darwinian anthropology, Lamarckian environmentalism, a Spencerian concern for the 'social organism', and a view of inheritance every bit as mad and imprecise as Darwin's own, provided the framework within which Olivier made his argument. And this almost brings us back to where we started – with Wallace's socialism growing out of his attempt to make a Darwinian case for racial equality. However, by the time Olivier wrote it was clear that a change was occurring in the relationship between socialism and Darwinism. Whereas Wallace had made his argument from the cutting edge of evolutionary science, Olivier had not even considered the latest research on inheritance by Weismann. A dual process was occurring that ensured that the distance between socialism and the latest detailed developments in Darwinian thinking was growing. On one side, science was becoming an ever more professional discipline and concomitantly detached from mainstream society. On the other, the burgeoning socialist movement was demanding ever clearer statements of the socialist creed.

10

Neither Liberalism nor Marxism

The Socialist Library provided clear and simple statements of the socialist creed, expressed in an organic and evolutionary language. In doing so it reflected the agenda of its editor. Vandervelde and Bernstein were consulted about which books to include but the Library was overwhelmingly the personal project of MacDonald.[1] He wrote three of the volumes and either commissioned or chose the rest for translation. The foreign authors were friends and colleagues from the Second International; the British ones were contacts MacDonald had made in twenty years of activity on the left. This eclectic collection fairly represented the eclectic nature of MacDonald's thought. The foreign socialists, with the exception of Jaurès, had received a formal scientific training. The British authors had none. MacDonald – the self-taught scientist – was equally comfortable with both and provided the Library with its intellectual rationale.

The claim in the *Prospectus* that the Library would not restrict itself to 'any particular school of thought' was disingenuous. The Library's readership was fed a constant diet of organic phraseology, to encourage them to think and express themselves in evolutionary terms. MacDonald's *Socialism and Society*, with its Spencer-inspired disquisitions on the similarities between society and an organism, was only the most obvious example of an approach that was found to differing extents in all of the Library's output. Vandervelde, for example, was equally prone to appeal to 'the organic character of the conception which serves as a basis for collectivist theories'.[2] And even more insidious than such explicit assertions of organicism was the frequency with which each of the authors in the Library portrayed specific social processes in organic terms. For MacDonald both imperialism and general elections were organic processes, Bernstein found co-operative stores 'organisms' and Snowden argued that the approach to temperance reform needed to be more organic than previously.[3]

A multitude of such examples could be drawn from any of the volumes in the Library. They were the product both of the isomorphic interaction between science and socialism in which many of the authors had developed their own political beliefs and a self-conscious attempt to encourage an organic and evolutionary mindset among their readers. MacDonald's ambition was to make a generation of socialists in his own image; men and women who would associate their socialist beliefs with Darwinism to such an extent that, as Margaret McMillan put it: 'Even those who forget the facts go home satisfied that a demand was made that is somehow in line with the forces of Evolution.'[4]

Moulding the membership

The success of the Socialist Library – and MacDonald's other forays into political education – in shaping the ILP membership to this mould is illustrated by the strategy MacDonald adopted to rebut revolutionary challenges when they arose. Questions concerning the meaning and methods of socialism, which had first surfaced during the 'Socialist Unity' debates of the 1890s, rumbled on throughout the first quarter of the twentieth century. At several crucial junctures – the 'Grayson affair' of 1909, the Syndicalist threat of 1912 and Bolshevism in 1919 – MacDonald was able to persuade and reassure the ILP membership by appealing to their evolutionary understanding in an explicitly organic language. His opponents, MacDonald would patiently explain, had mistaken society for an artificial machine, when it was in fact an evolving organism. That each time MacDonald succeeded in dissuading the ILP from a revolutionary path confirms his achievement in building 'a strong school of Revisionism' in Britain – at least in MacDonald's understanding of the term. There may have been little in the way of a direct assault on specific tenets of Marxist theory, but ILP members seem to have been overwhelmingly receptive to an evolutionary understanding of socialism expressed in an organic language. This suggests that an important element in any answer to the hoary old question 'Why was there no Marxism in Britain?' must be the bulwark built by an organic and evolutionary mindset among British socialists.[5]

This can be clearly seen in the 'Grayson affair' of 1909. Victor Grayson had won the Colne Valley by-election in 1907 as an 'independent socialist'. Thereafter he became both a beacon for the recalcitrant 'impossibilists' in the ILP and a thorn in the side of the party leadership. A glitch in the attempt of the NAC to reprimand Grayson at the 1909 annual conference provoked the 'Big Four' – Hardie, MacDonald, Snowden and Glasier – into resigning from the NAC. This mixture of pique and bullying had its desired effect, as conference repudiated its previous impertinence on what had in any case been a trivial point, and swung back behind the leadership. Of more interest and enduring importance than these machinations was MacDonald's valedictory chairman's address. Delivered as a barely disguised rebuke to Grayson, it was subsequently published as an official ILP pamphlet, *Socialism To-day* (1909). The speech is remarkable, not only for MacDonald's 'passionate and uncompromising' tone, but because MacDonald made his case in such explicitly organic and evolutionary terms.[6] Indeed, in the middle passages of the speech MacDonald discussed, in detail, the latest developments of the mutation theory of De Vries and the horticulture experiments of Burbank, before concluding that Grayson and the Social Democratic Federation lacked 'scientific faithfulness' and were 'foredoomed to failure' by their misunderstanding of the social organism.[7] It is significant that, faced with one of the most important speeches of his career and the need to win the conference round, MacDonald calculated that the best way to convince an ILP audience was to make a Darwinian case for socialism and to castigate his opponents as 'unscientific'.

The same approach saw off the threat of syndicalism in 1912. Syndicalism, MacDonald explained, proceeded from the wrong analogy and 'Error infinite creeps into our thoughts by false analogy.' By assuming a mechanical, rather than organic, relationship between the economy and society, syndicalism mistook class conflict – rather than social growth – for the source of socialism. This led it to adopt the erroneous method of the general strike.[8] This was even more misguided than Jacobinism. Whereas the insurrectionists briefly wielded the political power of society, the syndicalists – and this was what made syndicalism particularly galling for an evolutionist – simply proposed bringing society to a standstill, by paralysing the social organism. That this was unwise was illustrated by Metschnikoff's observations in biology. Paralysis in any organism always provoked a reaction from 'phagocytes or cells whose function is to render the organic life immune from disease'. The phagocytes of society, 'interests, prejudices, habits', would similarly react to paralysis when the 'social life' was threatened and the paralysis of the general strike would provoke a terrible reaction. Metschnikoff's physiological theories, MacDonald concluded, needed to be applied more thoroughly to sociology than had been done.[9] Thus, again, when faced with a challenge to his brand of socialism, MacDonald judged that the clinching arguments for ILP members lay in the scientific correctness of his socialism.

The decisive vindication, however, came when the ILP rejected affiliation to the Third International. This was a critical moment for the British left. If the ILP had joined the Third International, then communism would have gained a significant foothold and Britain might have experienced the same disabling and debilitating struggles between socialism and communism that so weakened the continental left and eased the path of fascism. The key to staving off this threat can be seen in the twelfth and final volume of the Socialist Library, *Parliament and Revolution*. Written in the immediate aftermath of war, revolution and the 'Coupon Election', this book, which Marquand judged '[i]n many ways . . . the most effective polemic [MacDonald] ever wrote', was a desperate plea to dissuade socialists from abandoning parliamentarianism.[10] To this end, MacDonald offered a parliamentary rendering of Kautsky's *Dictatorship of the Proletariat.*

MacDonald had long admired Kautsky – he had sought to procure his services for the *Socialist Review* – and the war had brought the two's positions closer.[11] Both had been isolated from their respective parties by an anti-militarist stance and the Bolshevik revolution had shifted the lines of division within international socialism. The real point of congruence, and the real power of *Parliament and Revolution*, lay in the language MacDonald used. As in Kautsky's *Dictatorship of the Proletariat*, the crux of MacDonald's critique was that the Bolsheviks were following the utopian path of 1792 and had failed to realize that the evolutionary alternative of a 'tortoise pace' would ultimately prove more successful than the fitful activities of a Jacobin minority. MacDonald even joined with Kautsky in accusing the Bolsheviks and their followers of 'discarding the historical, scientific method of Marx' and forgetting 'that no people can overcome the obstacles offered by successive phases of their development by a jump or a legal enactment'. Once it was accepted that society was an organism, it became obvious that planning a revolution, attempting to impose a programme by force, or hastening social change was absurd. Political power was not separate from society, as the Bolsheviks imagined, nor was society a rigid and unresponsive mechanism. A socialist could not fight nature; his spirit had to be 'historical and not cataclysmic'. MacDonald's *coup de grâce* was to point out that, whilst the followers of the Russian method were 'specially fond of calling themselves "scientific"', this adjective belonged exclusively to organic and evolutionary socialists.[12]

Post-war political education

After the First World War, a new generation of socialists were 'clamouring for classes' and 'keenly enthusiastic [for] devouring literature'.[13] The Socialist Library set the template for their political education, but a generational shift made them less receptive to the message. The Liverpool delegate to the ILP's 1919 conference who demanded literature of an 'increasingly Marxian' character, 'in order to create Socialists, class-conscious rebels', was to be sorely disappointed.[14] Instead a new Information Committee, chaired by MacDonald and charged with providing a 'great propaganda by leaflets, pamphlets and books', was established.[15] Under its auspices two new book series emerged. The 'Social Studies Series' was the direct successor to the Socialist Library. Its opening volume, MacDonald's *Parliament and Democracy* (1921), was the companion volume to the Library's last number, *Parliament and*

Revolution.[16] More innovative was the 'ILP Study Courses' series, launched in 1920.

This series is a goldmine for anyone interested in the reception and diffusion of socialist ideas. Each book provided notes for lecturers and class leaders, explaining how they should structure discussions around different aspects of socialist theory and ILP policy, with accompanying notes for further reading. Although we cannot know how closely these were followed, they clearly provide an excellent source for understanding how socialism was discussed and understood by the lay membership. And they confirm the continuing desire of the ILP leadership to make socialists who thought in evolutionary terms and expressed themselves in an organic language. Not only were texts from the Socialist Library frequently included in the recommended reading sections – for example, Mary Agnes Hamilton's *The Principles of Socialism* (1921) recommended MacDonald's *Socialism and Society* as the best guide to her subject – but the overall message of the series was the same.[17] MacDonald's opening volume, *The History of the ILP* (1921), contained an injunction for class leaders to trace 'the steady evolution of [the] idea' of socialism, up to the fully 'scientific' ILP, and 'If the leader can, he should refer to the advance in scientific and sociological thought since Marx's time, pointing out how the ILP position . . . far from being out of date, is in accord with modern thought.'[18]

A non-liberal lineage

Reclaiming the label 'scientific' from Marxist opponents was one thing, but separating ILP socialism from liberalism was not so easy. Indeed, some evidence would suggest that MacDonald was not that keen on even trying. If he was desperate to mark off the ILP from Marxist socialism, he seemed far less bothered about blurring the boundaries with liberalism. In 'The ILP Programme' of 1899, Hardie and MacDonald located their organization in the traditions of British radicalism and liberalism, and throughout the Edwardian period MacDonald presented the Labour Party as 'the child in direct line of succession' to the liberals. Liberalism, he often proclaimed, was the caterpillar and socialism the butterfly destined to succeed it.[19] The rise and fall of political parties, moreover, was held to accord with the laws that governed all organisms and because there were no 'gulfs' in evolution Labour would smoothly succeed Liberal. Just as

> [l]ower forms merge into higher forms, one species with another, the vegetable into the animal kingdom; in human history one epoch slides into another . . . Socialism, the stage which follows Liberalism, retains everything of value in Liberalism by virtue of its being the hereditary heir to Liberalism.[20]

It is tempting to read such statements as unproblematic evidence of the 'currents of radicalism' flowing between late-nineteenth-century liberalism and socialism. And it is a temptation to which many historians, most notably Reid and Biagini, have succumbed. Inverting Sir William Harcourt's famous judgement of late-nineteenth-century politics that 'We are all Socialists now', the 'currents of radicalism' school has instead declared 'They were all liberals really'.[21] It is an interpretation that is particularly seductive in the case of MacDonald as it helps dispense with the messy business of explaining the betrayal of 1931. But, as we argued in Chapter 5, despite an overlap between MacDonald's socialism and Edwardian liberalism – especially the New Liberalism – there were important points of divergence that provide evidence of a distinctive philosophical position.

Before rushing to judgement on MacDonald's statements about liberalism, it also worth pausing to consider his very similar remarks about Marxism and other early-nineteenth-century forms of socialism. Here MacDonald's attitude was equally genealogical. Just as he found liberalism passing into the ILP and the Labour Party, so MacDonald argued that socialist predecessors, from Thomas Paine through the Owenites and Chartists, and even Marx and Engels, were the forefathers of his own organizations.[22] In this light, MacDonald's comments on liberalism assume a different meaning. Gray and others have a point in criticizing Reid and Biagini for failing to appreciate that their evidence for continuity can also be read critically as a 'contentious attempt to claim a tradition, rather than evidence that one existed'.[23] Edwardian socialists were engaged in a bitter struggle to portray themselves as the heirs to both the liberal and socialist traditions, as a means of usurping the political mantle of their rivals.

In the case of liberalism, the claim for a common heritage was tied to the assertion that liberalism was 'a creed of the past' and that the Liberal Party was incompetent to guard its own traditions and principles.[24] MacDonald taunted Hobhouse for not being able to 'get out of the fact that his Liberal system of thought has passed into the Socialist system of thought'. At every point in his review of *What is Liberalism?*, MacDonald declared, 'we have it borne in upon us that pure Liberalism is a creed of the past, surviving in the present like a strain in a mixed generation'.[25] This suggestion of heredity was not accidental. MacDonald's comments on both liberalism and Marxism clearly served the same dual end – claiming the inheritance of a political competitor and sidelining that rival as anachronistic.

The danger with Gray's interpretation, however, is that it implicitly suggests an instrumental use of an organic and evolutionary language, which rather underestimates the extent to which MacDonald and his contemporaries really believed their own analysis. It is no coincidence that those who sought to rediscover the 'early English socialists' – the Webbs,

MacDonald, H. S. Foxwell, for example – were also impressed by evolutionary theory. For them the socialist movement had to be, as MacDonald put it, 'a historical growth and not a mathematical discovery', and this meant finding the small beginnings from which it had grown.[26] This was not a peculiarly British phenomenon. The same interpretation can be found in the writings of continental socialists of this period. Just as MacDonald traced the evolution of British socialism from Thomas Paine, through the Owenites and Chartists, up to the Independent Labour Party, so Jaurès found French socialism evolving from the revolution through Baeuf, Fourier, Saint-Simon and Proudhon.[27] The point for MacDonald was to assert his own socialism as an evolutionary product that was best adapted to the contemporary environment. The same claim was the essence of Revisionism. Both MacDonald's socialism and Revisionism were constructed in a Darwinian discursive space and derived a heuristic impetus from the organic and evolutionary language in which they were expressed. This provided them with a discrete identity that marked them off from liberalism on one side and Marxism on the other.

Conclusion

Historians of science have long been familiar with the concept of a *constitutive metaphor*. This is a metaphor that 'constitutes, at least for a time, an irreplaceable part of the linguistic machinery of scientific theory'. Where no literal paraphrase is known, usually at the inception of a new theory, scientists depend upon this metaphor to develop their thought. It is, in turn, articulated, probed and extended by other scientists, to the extent that it ceases to be simply a metaphor and enters the process of thought itself, performing the vital cognitive function of framing the discursive patterns of the theory. Eventually, as the new science becomes established, and it is increasingly possible to assign more literal names to processes, so the importance of the metaphor declines.[1] A similar pattern can be seen in social thought. Late-nineteenth-century physics, for example, has been identified as providing neo-classical economics with its constitutive metaphor, initially helping to determine and frame the content of economic theory until such a point was reached where the metaphor could be dispensed with.[2]

It has been our contention that socialism too, in the period between 1859 and 1914, had its own constitutive metaphor loosely drawn from Darwinism. We have argued that the undeniable growth in the use of an organic and evolutionary language on the left was not simply a 'sign of the times' – mirroring the passing of Carlyle's 'machine age' into Spencer's 'Social Organism' – or an instrumental deployment of a convenient argument. Rather, it signalled a stage in socialism's own development where 'no literal paraphrase' could be found for the views socialists wanted to present and where no historical or foreign precedence could be called upon to justify their case. This mattered because as socialists articulated, probed and extended the organic and evolutionary metaphor, and expressed their argument in its language, so it entered into their very processes of thought. In particular, Revisionism on the continent, and MacDonald's version of socialism in Britain, are inconceivable without Darwinism as their constitutive metaphor.

This relationship with Darwinism brought considerable benefits to the left. It provided the discursive space in which socialism could develop and flourish; one, moreover, that for a time connected the left to the most vibrant and advanced scientific thought. And in opposition, which with a few minor and brief exceptions was where European socialists resided before 1914, the socialist exegesis of Darwinism bred an unquenchable optimism. The direction of organic growth was towards an ever more socialistic state and, though the odd crisis might lead to a brief period of regression, the future was assured. But constitutive metaphors are only temporary and transient. Once a science has achieved a certain level of growth, the weaknesses that originally necessitated its use should be overcome. At this point the metaphor falls away as smoothly as the cocoon from the body of the butterfly, to reveal a new, independent and self-sufficient body of thought.

Unfortunately the transition was not quite as smooth in the case of socialism. As the movement grew in size and confidence, and began to experience the thrill of power, so socialism as an ideology should have begun to feel its own strength and break free of the discursive space in which it had been nurtured. There would have been no need to eschew organicism and evolutionism – they would merely have fallen back into being one of many linguistic devices available in the armoury of the socialist thinker. In their place an argument for socialism built upon human beings' command of their own cultural environment might have been expected to arise. And indeed, to a limited extent, this process did begin to occur.

Even in the period 1859 to 1914 the character of the interaction between socialism and Darwinism had begun to undergo a subtle change. From the cutting-edge science of Wallace, through the eclectic populist interpretation of MacDonald, to a shallower understanding in the political education activities of the Socialist Library, a changing pattern can be discerned in the left's relationship with Darwinism. Over time Darwinism declined as a source of innovation and development on the left and increasingly became a means to express a settled doctrine. There were no distinct stages and no rigid sequence in this process. There was no three-tiered hierarchy of intellectuals, politicians and activists, and therefore no ideological trickledown effect. Lay activists had direct access to Darwin's ideas from 1859 and many socialists received their schooling in evolutionary ideas from the mechanics' institutes of the 1880s and 1890s.[3] But though it remained essential to the way socialists thought and acted, at some point in the Edwardian period it seems reasonable to conclude that the Darwinian metaphor had ceased to contribute to the development of socialist arguments. In part this was the product of the professionalization of science, which meant that a self-educated maverick like Wallace would never again trouble the frontiers of knowledge.[4] In part it reflected the growing self-sufficiency of the socialist movement. The overall effect was to render Darwinism, as a heuristic structure for the left, increasingly moribund. There was less and less room for innovation. What had begun life as a supple and evolving system of thought – that could be developed in imaginative ways – solidified into a case that contained, rather than stimulated, socialist thought. At that point the skin should have been shed.

That is precisely what the ILP attempted in the mid-1920s. MacDonald's post-war political education campaigns had been far less successful than the Socialist Library. And at the 1926 conference, the party ignored MacDonald's advice and adopted a policy document, *Socialism in Our Time*, which rejected 'the deadening idea that Socialism can only be established by slow gradualism over generations of time'.[5] The following year MacDonald resigned from the party he had done so much to shape, complaining that it was no longer 'a good Socialist body'.[6] One should avoid overstating the ideological significance of these two events – MacDonald's disenchantment was personal as much as political and *Socialism in Our Time* was opposed by 'important sections of the party'.[7] Nonetheless, taken together they confirm a definite change. A new generation of ILP activists were demanding what MacDonald refused to provide – 'purpose and vitality to our propaganda for socialism, which no theory based on the ripening of events can give'.[8] MacDonald had shown 'disastrously little interest in the ideas of young men'.[9] He could not fully understand why this new cohort were unmoved by his evolutionary arguments and lacked the reverence of his own cohort for all things Darwinian. If these developments in the ILP had been a straw in the wind for a recasting of socialist thought – with organicism and evolutionism reduced to a position of less prominence – then the disaster of the second Labour government might have been avoided.

Unfortunately, MacDonald's loss of influence in the ILP was more than compensated for by his intellectual ascendancy within the Labour Party.[10] This ensured that the government which came to power seventy years after the publication of the *Origin*, and ended in ignominious failure sixty years after the publication of the *Descent*, was headed by a man dependent upon Darwinism as the constitutive metaphor in his thought. This fact ought to alert us both to the lengthy time lags involved in the dissemination of science and to the weaknesses inherent in the Darwinian heuristic for the left. As with any discursive space, Darwinism imposed boundaries and set limitations. Two, in particular, proved problematic. The first was the active discouragement to the making of detailed plans and programmes, which a faith in an organic and evolutionary development engendered. The second was a mindset that had almost gone into denial about the possibility of catastrophe. Together they left MacDonald and his government with a woefully inadequate understanding of the crash of the capitalist system that engulfed them soon after they assumed office.

As the harsh conditions of the Great Depression took hold, MacDonald clung ever more tightly to the evolutionary faith that had sustained him in his youth. And the Labour Party followed him. At the 1930 party conference, Oswald Mosely's demands for an active, interventionist economic strategy were rebutted in a MacDonald speech proclaiming optimism in the future evolution of society. But by this time it was obvious that the system of thought MacDonald had spent a lifetime developing had denuded the left of the ability to understand cataclysms and had actively encouraged the failure to develop alternative policies. All too quickly this brought its inevitable denouement in the fatal collapse into economic orthodoxy that brought down the government.

A connection with the collapse of the second Labour government has not been made in order to condemn the use of Darwinism as a constitutive metaphor in the period up to 1914. That would be strange indeed, given our argument that there was little conscious choice involved. The notion of a Darwinian left was not inherently flawed – at least in the period before 1914. The problem arose less from a linguistic determinism than from a failure to remove a leadership that was trapped in its dependence upon a metaphor that had passed its sell-by-date. For this the party was to blame as much as MacDonald. Both clung to the comfort blanket of their constitutive metaphor long after a new generation of leaders, ready to develop socialism as an established ideology, ought to have arisen.

That it took a mixture of betrayal and collapse, for the party to recognize that the intellectual swaddling clothes of Darwinism could be cast aside, was not exclusively MacDonald's fault. When a new leadership did eventually arise, under Clement Attlee, it was built on a recognition that MacDonald's greatest failing had been to place too much faith in gradual evolution and to give too much emphasis to organic unity. Attlee, it goes without saying, was no revolutionary. As a graduate of the ILP, he echoed MacDonald's evolutionary account of the history of the Labour Party, and stressed its anti-utopian credentials.[11] He even fretted over the possibility that the 'revulsion from MacDonaldism' might cause 'the Party to lean rather too far towards a catastrophic view of progress and to emphasise unduly the conditions of crisis which were being experienced, and to underestimate the recuperative powers of the Capitalist system'.[12] But although organic and evolutionary metaphors remained a resource to which Attlee often returned, they no longer enjoyed a controlling position.

This offers an important lesson to those calling for a new Darwinian left. The left needs Darwinism, but Darwinism in its proper place – not as a master discourse or a constitutive metaphor. Of course, there are no direct parallels in history. Both socialism and Darwinism have themselves evolved significantly in the past seventy years. But to follow Singer's advice, and once again make Darwinism the discursive space in which the socialism is constructed, would have equally distorting and harmful consequences. Most obviously it would strip the left of its belief in the capacity of humans to change themselves and society. Singer's key point is that the left needs to focus on what is static and unchanging in human nature rather than on what is malleable and evolving. The irony that statis is invoked in the name of Darwinism is lost on Singer, who sees himself reacting to a traditional left-wing denial of human nature. This, as we have seen, is mistaken, and it is about time the left stopped trying to build its future on a disavowal of its past.

The left has never denied the concept of human nature. What it has done – from Marx to Chomsky – is to suggest that the innate parameters of human nature should be drawn widely enough to permit a large degree of change; not in defiance of Darwinism but because historical evidence points to the potential for changes in the economic and social structure to precipitate fundamental changes in patterns of social behaviour. This is the key to removing the main economic root of social conflict and radically reducing inequality. Mankind may be constrained by human nature but the past century of change in the so-called 'nature' of women has shown us how little of our character that 'nature' constitutes. Similarly, in 1906 Sydney Olivier noted how many racial inequalities that had been thought eternal only a hundred years previous had been overcome 'by ignoring the obvious; by refusing to accept as conclusive the differences and the disabilities; by believing in the identities, the flashes of response and promise'.[13] This message, that humans can shape and reshape the cultural space they inhabit, is the counsel of hope upon which the left must be built.

Notes

Introduction: Myths and Misunderstandings

1. See, for example, I. Berlin, *Karl Marx* (London, 1963), pp. 247–8; S. Avineri, 'From hoax to dogma – a footnote on Marx and Darwin', *Encounter* (March 1967), pp. 30–2; B. Ollmann, *Alienation: Marx's conception of man in capitalist society* (Cambridge, 1971), p. 53; D. McLellan, *Karl Marx: His life and thought* (London, 1973), p. 424; V. Gerratana, 'Marx and Darwin', *New Left Review*, 35 (1974), pp. 60–82. For a flavour of how the myth was eventually disproved, see L. S. Feuer, 'Is the "Darwin–Marx Correspondence" authentic?', *Annals of Science*, 32 (1975), pp. 1–12; R. Colp Jr, 'The contacts of Charles Darwin with Edward Aveling and Karl Marx', *Annals of Science*, 33 (1976), pp. 387–94; M. A. Fay, 'Did Marx offer to dedicate *Capital* to Darwin? A reassessment of the evidence', *Journal of the History of Ideas*, 39 (1978), pp. 133–46; L. S. Feuer, 'The case of the Marx–Darwin letter', *Encounter* (October 1978), pp. 62–78.

2. *New Century Review* (March–April 1897), pp. 232–43.

3. K. Marx and F. Engels, *Collected Works*, vol. 26 (London, 1990), p. 517.

4. See T. Benton, 'Social Darwinism and socialist Darwinism in Germany: 1860–1900', *Rivista di filosofia*, 73 (1982), pp. 79–112.

5. Darwin to Dr Scherzer, 26 December 1879, in F. Darwin (ed.), *The Life and Letters of Charles Darwin, including an Autobiographical Chapter* (London, 1887), vol. III, pp. 236–7. For the broader context of this debate, see R. Weikart, 'The origins of social Darwinism in Germany, 1859–1895', *Journal of the History of Ideas*, 54 (1993), pp. 569–88.

6. See, for example, Marx to Engels, 7 December 1867, *Collected Works*, vol. 42, p. 494.

7. J. Rifkin, 'This is the age of biology', *Guardian*, 28 July 2001; see also L. Arnhart, 'The new Darwinian naturalism in politial theory', *American Political Science Review*, 89 (1995), pp. 389–400.

8. P. Singer, *A Darwinian Left: Politics, evolution and cooperation* (London, 1999), p. 57.

9. R. Thornhill and C. T. Palmer, *A Natural History of Rape: Biological bases of sexual coercion* (Cambridge, Mass., 2000). See also D. M. Buss, *Evolutionary Psychology: The new science of the mind* (Boston, 1999).

10. Singer, *Darwinian Left*, p. 6.

11. See also in the 'Darwinism Today' series, K. Browne, *Divided Labours: An evolutionary view of women at work* (London, 1998). For a partial corrective to such nonsense, see L. Rogers, *Sexing the Brain* (London, 1999).

12. Singer, *Darwinian Left*, p. 62.

13. Ibid., p. 5.

14. Ibid., p. 62.

15. Ibid., p. 16.

16. Ibid., pp. 39, 36.
17. Ibid., p. 5.
18. Ibid., pp. 19–20.
19. A. Besant, *Why I am a Socialist* (London, 1886), p. 2.
20. R. Cooter, *The Cultural Meaning of Popular Science* (Cambridge, 1985), p. 2.
21. A. Desmond, 'Artisan resistance and evolution in Britain, 1819–1848', *Osiris*, 3 (1987), p. 89.
22. R. Cooter, 'The power of the body: the early nineteenth century', in B. Barnes and S. Shapin (eds), *Natural Order: Historical studies of scientific culture* (London, 1979); F. E. Manuel, 'From equality to organicism', *Journal of the History of Ideas*, 17 (1956), pp. 54–69.
23. D. Stack, 'William Lovett and the National Association for the Political and Social Improvement of the People', *Historical Journal*, 42 (1999), pp. 1027–50.
24. C. Darwin, *The Origin of Species by means of natural selection or the preservation of favoured races in the struggle for life* (Harmondsworth, 1985), p. 263.
25. For the construction of the narrow definition of Darwinism, see J. Moore, 'Deconstructing Darwinism: the politics of evolution in the 1860s', *Journal of the History of Biology*, 24 (1991), pp. 353–408.
26. M. Pittenger, *American Socialists and Evolutionary Thought, 1870–1920* (Madison, Wisc., 1993); G. Jones, *Social Darwinism and English Thought: The interaction between biological and social theory* (Brighton, 1980); D. P. Todes, *Darwin without Malthus: The struggle for existence in Russian evolutionary thought* (Oxford, 1989).
27. For the latest survey of social Darwinism, see G. Claeys, 'The "survival of the fittest" and the origins of social Darwinism', *Journal of the History of Ideas*, 61 (2000), pp. 223–40.
28. R. Hofstadter, *Social Darwinism and American Thought* (New York, 1959).
29. E. Biagini and A. Reid (eds), *Currents of Radicalism: Popular radicalism, organised labour and party politics in Britain, 1850–1914* (Cambridge, 1991). For an introduction to the debate over 'continuity' in British radicalism and socialism, see R. McWilliam, *Popular Politics in Nineteenth-century England* (London, 1998).
30. M. Taylor, 'The Six Points: Chartism and the reform of Parliament', in O. Ashton, R. Fyson and S. Roberts (eds), *The Chartist Legacy* (Woodbridge, 1999), pp. 1–23; F. Trentmann, 'The strange death of free trade: the erosion of "liberal consensus" in Great Britain, *c.*1903–1932', in E. Biagini (ed.), *Citizenship and Community: Liberals, radicals and collective identities in the British Isles, 1865–1931* (Cambridge, 1996), pp. 232–3.
31. G. Stedman Jones, 'The determinist fix: some obstacles to the further development of the linguistic approach to history in the 1990s', *History Workshop Journal*, 42 (1996), pp. 19–35.

Chapter 1 Darwin's Challenge

1. C. Darwin, *The Descent of Man, and selection in relation to sex* (London, 1871), vol. II, p. 388fn.
2. D. Stack, *Nature and Artifice: The life and thought of Thomas Hodgskin, 1787–1869* (London, 1998), pp. 224–32.

3. Marx to Engels, 19 December 1860, in K. Marx and F. Engels, *Collected Works*, vol. 41 (London, 1985), p. 232.

4. See L. Jordanova, *Lamarck* (Oxford, 1984).

5. N. Garfinkle, 'Science and religion in England, 1790–1800: the critical response to the work of Erasmus Darwin', *Journal of the History of Ideas* (1955), pp. 376–88.

6. For Paley, see D. L. Le Mahieu, *The Mind of William Paley* (London, 1976).

7. See A. Desmond, 'Artisan resistance and evolution in Britain, 1819–1848', *Osirisis*, 3 (1987), pp. 77–110.

8. See J. Secord, *Victorian Sensation: The extraordinary publication, reception and secret authorship of Vestiges of the Natural History of Creation* (Chicago, 2000).

9. This parallel is drawn in Secord's 'Introduction' to R. Chambers, *Vestiges of the Natural History of Creation and Other Evolutionary Writings* (Chicago, 1994).

10. Darwin to Galton, 28 May 1873, in F. Darwin (ed.), *The Life and Letters of Charles Darwin, including an Autobiographical Chapter* (London, 1887), vol. III, p. 178.

11. A. Desmond and J. Moore, *Darwin* (London, 1992), pp. 540–1; R. Weikart, 'A recently discovered Darwin letter on social Darwinism', *Isis*, 86 (1995), pp. 609–11.

12. A. Desmond, *The Politics of Evolution: Morphology, medicine, and reform in radical London* (Chicago, 1989), especially pp. 398–414.

13. E. Ferri, *Socialism and Positive Science (Darwin–Spencer–Marx)* (London, 1905), pp. 1–7.

14. See P. H. Barrett, P. J. Gastrey, S. Herbert, D. Kohn and S. Smith (eds), *Charles Darwin's Notebooks, 1836–1844* (Cambridge, 1987).

15. Desmond and Moore, *Darwin*, p. xvi.

16. From the opening paragraph of Darwin's paper to the Linnean Society, 1 July 1858. Reproduced in J. L. Brooks, *Just Before the Origin: Alfred Russel Wallace's Theory of Evolution* (New York, 1984), p. 259. See R. Young, 'Malthus and the evolutionists: the common context of biological and social theory', *Past and Present*, 43 (1969), pp. 109–45.

17. See A. La Vergata, 'Images of Darwin: a historiographic overview' in D. Kohn (ed.), *The Darwinian Heritage* (Princeton, 1985), pp. 901–72, especially pp. 953–8; E. Manier, *The Young Darwin and his Cultural Circle: A study of influences which helped shape the language and logic of the first draft of the theory of natural selection* (Boston, 1978).

18. The 'modern synthesis' is the name given to the reformulation of Darwinism, incorporating genetics, in the period 1936–47. For a discussion of the relationship between Darwin's theory and modern Darwinism, see E. Sober, *The Nature of Selection* (Cambridge, 1985), ch. 6. See also J. Huxley, *Evolution: The modern synthesis*, Harvard, 1980).

19. R. L. Meek, *Marx and Engels on Malthus: Selections from the writings of Marx and Engels dealing with the views of Thomas Robert Malthus* (London, 1953).

20. R. Dean, 'Owenism and the Malthusian population question, 1815–1835', *History of Political Economy*, 27 (1995), pp. 579–97.

21. J. Robertson, *Socialism and Malthusianism* (London, 1885).

22. F. Engels, 'Outlines of a Critique of Political Economy', in W. O. Hen-

derson (ed.), *Selected Writings* (Harmondsworth, 1967), p. 150. And this is still often the view of the left: see Stephen J. Gould – 'Darwin may have cribbed the idea of natural selection from economics, but it may still be right' – quoted in La Vergata, 'Images of Darwin'.

23. See A. McLaren, *Birth Control in Nineteenth Century England* (London, 1978).

24. That Darwin himself was often unable to resolve all the issues his ideas raised can be seen in E. Mayr, 'Darwin's five theories of evolution', in D. Kohn (ed.), *The Darwinian Heritage* (Princeton, 1985).

25. See C. Darwin, *The Variation of Animals and Plants under Domestication* (London, 1868).

26. In 1876 Darwin was prepared to admit that 'the greatest error which I have committed, has been not allowing sufficient weight to the direct action of the environment, i.e. food, climate etc., independently of natural selection'. C. Darwin to M. Wagner, 13 October 1876, in F. Darwin (ed.), *Charles Darwin* (London, 1902), p. 278.

27. P. J. Bowler, *Evolution: The history of an idea* (Berkeley, 1989), pp. 171, 190, 210.

28. M. Hawkins, *Social Darwinism in European and American Thought, 1860–1945* (Cambridge, 1997), pp. 26–8.

29. Bowler, *Evolution*, pp. 232–3.

30. See G. Stocking, *Victorian Anthropology* (New York, 1987), and G. Stocking, *Race, Culture and Evolution: Essays in the history of anthropology* (New York, 1968).

31. P. J. Bowler, 'The changing meaning of "evolution"', *Journal of the History of Ideas*, 36 (1975), pp. 95–114.

32. O. Chadwick, *The Secularisation of the European Mind in the Nineteenth Century* (Cambridge, 1975), p. 161.

33. See the quotes from Whewell and Bacon on the title page. C. Darwin, *The Origin of Species by means of natural selection or the preservation of favoured races in the struggle for life* (London, 1985).

34. J. S. Mill, 'On nature', in his *Three Essays on Religion* (London, 1874).

35. On Darwin's religious views, see M. Mandelbaum, 'Darwin's religious views', *Journal of the History of Ideas*, 19 (1958), pp. 363–78, and J. Moore, *The Darwin Legend* (London, 1994).

36. J. R. Lucas, 'Wilberforce and Huxley: a legendary encounter', *Historical Journal*, 32 (1979), pp. 313–30; J. Moore, 'Freethought, secularism, agnosticism: the case of Charles Darwin', in G. Parsons (ed.), *Religion in Victorian Britain*, vol. 1: *Traditions* (Manchester, 1988), pp. 274–319; D. Osvopat, *The Development of Darwin's Theory: Natural history, natural theology and natural selection, 1838–1859* (Cambridge, 1981).

37. On this see M. Ruse, *The Darwinian Paradigm* (London, 1989), pp. 34–54.

38. Darwin, *Descent*, vol. I, pp. 163–4, although cf. pp. 98–9.

39. Ibid., p. 166.

40. C. Darwin, *The formation of vegetable mould through the action of worms, with observations on their habits* (London, 1881).

41. See J. Maynard Smith, 'Sexual selection', in S. A. Barnett (ed.), *A Century of Darwin* (London, 1962), pp. 231–44.

42. Darwin, *Descent*, vol. II, pp. 326–9.

43. Ibid., vol. II, p. 240.

Chapter 2 Alfred Russel Wallace

1. A. Desmond and J. Moore, *Darwin* (London, 1992), p. 669. The relative neglect of Wallace will soon be remedied. A new biography is due this year, M. Shermer, *Heretic Scientist: The life and science of Alfred Russel Wallace* (Oxford, 2002), and James Moore is also set to produce a substantial study on Wallace. See also P. Raby, *Alfred Russel Wallace: A life* (London, 2001).

2. On Wallace, see H. Lewis McKinney, *Wallace and Natural Selection* (New Haven, 1972); M. J. Kottler, 'Charles Darwin and Alfred Russel Wallace: two decades of debate over natural selection', in D. Kohn (ed.), *The Darwinian Heritage* (Princeton, 1985), pp. 367–432; R. Smith, 'Alfred Russel Wallace: philosophy of nature and man', *British Journal for the History of Science*, 6 (1972), pp. 177–99.

3. A. R. Wallace, *My Life: A record of events and opinions*, vol. 1 (London, 1905), p. 234.

4. See B. G. Beddall, *Wallace and Bates in the Tropics: An introduction to the theory of natural selection* (London, 1969).

5. However, as with Darwin's 'eureka-moment', recent scholarship has emphasized the long, cumulative process involved. See J. Moore, 'Wallace's Malthusian moment: the common context revisited', in B. Lightman (ed.), *Victorian Science in Context* (Chicago, 1997), pp. 290–311.

6. Desmond and Moore, *Darwin*, p. 469.

7. See B. G. Beddall, 'Wallace, Darwin, and the theory of natural selection: a study in the development of ideas and attitudes', *Journal of the History of Biology*, 1 (1968), pp. 261–323; H. Lewis McKinney, *Wallace and Natural Selection* (New Haven, 1972), pp. 131–46; Desmond and Moore, *Darwin*, pp. 466–70.

8. C. Darwin, *The Origin of Species by means of natural selection or the preservation of favoured races in the struggle for life* (London, 1985), p. 458.

9. J. R. Schwartz, 'Darwin, Wallace, and the *Descent of Man*', *Journal of the History of Biology*, 17 (1984), pp. 271–89, 271–2.

10. A. R. Wallace, 'Sir Charles Lyell on geological climates and the origin of species', *Quarterly Review*, 1869 (126), pp. 359–94.

11. Desmond and Moore, *Darwin*, p. 569.

12. Reproduced in A. R. Wallace, *Contributions to the Theory of Natural Selection* (London, 1875), pp. 332–71.

13. See R. Rainger, 'Race, politics and science: the Anthropological Society of London in the 1860s', *Victorian Studies*, 22 (1978), pp. 51–70; N. Stepan, *The Idea of Race in Science: Britain, 1800–1960* (London, 1982); D. Lorimer, 'Theoretical racism in late Victorian anthropology', *Victorian Studies*, 31 (1988), pp. 405–30.

14. C. Darwin, *The Descent of Man, and selection in relation to sex* (London, 1871), vol. I, p. 235; on the cultural politicking around this paper, see E. Richards, 'The "Moral Anatomy" of Robert Knox: the interplay between biological and social thought in Victorian scientific naturalism', *Journal of the History of Biology*, 22 (1989), pp. 373–436.

15. A. R. Wallace, 'The development of human races under the law of natural selection', in Wallace, *Contributions*, pp. 303–31, 306–7.

16. Ibid., p. 325.

17. Ibid., p. 312.

18. Ibid., p. 318.

19. Ibid., pp. 329–31.

20. A. R. Wallace, 'The neglect of phrenology', in his *The Wonderful Century* (London, 1898), pp. 159–83. 'Wallace's belief in phrenology allowed him to discard natural selection as an influence in the human body when he wrote his paper in 1864' (Schwartz, 'Darwin, Wallace and the *Descent*', p. 283).

21. A. R. Wallace, 'The limits of natural selection as applied to man', in Wallace, *Contributions*, pp. 332–59, 344–9.

22. Ibid., p. 343.

23. A. R. Wallace, *Darwinism: An exposition of the theory of natural selection with some of its applications* (London, 1890), p. viii. Romanes characterized Wallace as a 'neo-Darwinian' for 'seeking to out-Darwin Darwin by assigning an exclusive prerogative to natural selection' (G. J. Romanes, *Darwin and After Darwin: Post-Darwinian questions – heredity and utility* (Chicago, 1895), vol. 2, pp. 12–13).

24. Ibid., p. 444.

25. Ibid., pp. 445–76, 'Darwinism applied to man'.

26. L. Barrow, *Independent Spirits: Spiritualism and English plebeians 1850–1910* (London, 1986).

27. Wallace, *Darwinism*, p. 478.

28. Wallace, 'The development', p. 317.

29. W. R. Greg, 'On the failure of "natural selection" in the case of man', *Frasers Magazine*, 78 (1868), pp. 353–62. See also the anonymous reply in *Quarterly Journal of Science*, 6 (1869), p. 152.

30. B. Semmel, *The Governor Eyre Controversy* (London, 1962).

31. Desmond and Moore, *Darwin*, pp. 569–70.

32. Wallace, *Darwinism*, pp. 474–6.

33. L. P. Curtis, *Apes and Angels: The Irishman in Victorian caricature* (Newton Abbot, 1971).

34. See, for example, the following comment from Darwin: 'For my part I would as soon be descended from that heroic little monkey, who braved his dreaded enemy in order to save the life of his keeper; or from that old baboon, who, descending from the mountains, carried away in triumph his young comrade from a crowd of astonished dogs – as from a savage who delights to torture his enemies, offers up blood sacrifices, practises infanticide without remorse, treats his wives like slaves, knows no decency, and is haunted by the grossest superstitions' (Darwin, *Descent*, vol. II, pp. 404–5).

35. Darwin, *Descent*, vol. I, p. 240.

36. See, for example, T. H. Huxley, 'Emancipation black and white', in his *Lay Sermons* (London, 1870).

37. J. R. Durant, 'Scientific naturalism and social reform in the thought of Alfred Russel Wallace', *British Journal for the History of Science*, 12 (1979), pp. 31–58, 43.

38. R. Young, 'Non-scientific factors in the Darwinian debate', *Actes XII Congrès International History of Science*, 8 (1968), pp. 221–6.

39. A. R. Wallace, *Social Environment and Moral Progress* (London, 1913), pp. 93–102.

40. Ibid., pp. 133–49.

41. 'The truly surprising feature of Wallace's long career is not that he became involved in so many cranky or pseudo-scientific causes, but rather that, through it all he clung on to a view of man and society which was still, in essence, naturalistic' (Durant, 'Scientific naturalism', p. 53).

Chapter 3 From Radicalism to Socialism

1. G. Stedman Jones, 'Rethinking Chartism', in his *Languages of Class: Studies in English working-class history, 1832–1982* (Cambridge, 1983), p. 105.
2. See D. Stack, *Nature and Artifice: The life and thought of Thomas Hodgskin, 1787–1869* (London, 1998), ch. 1.
3. This is a point of divergence between historians of political thought and historians of science. Whereas the former have concentrated on the mainstream providentialism of radicals, historians of science, especially Desmond, have been drawn to the atheistic fringe. Cf. Jones, 'Rethinking' with A. Desmond, 'Artisan resistance and evolution in Britain, 1819–1847', *Osiris*, 3 (1987), pp. 77–110.
4. P. J. Bowler, *The Non-Darwinian Revolution: Reinterpreting a historical myth* (Baltimore, 1988); I. W. Burrow, *Evolution and Society: A study in Victorian social theory* (Cambridge, 1966).
5. P. d'Arby Jones, *Henry George and British Socialism* (London, 1991), p. 1; E. P. Lawrence, 'Uneasy alliance: the reception of Henry George by British socialists in the '80s', *American Journal of Economics and Sociology*, II (1951), pp. 61–74.
6. Just like Hodgskin he dealt in turn with the wages fund, Malthus and property rights. See M. Blaug, 'Henry George: rebel with a cause', *European Journal of the History of Economic Thought*, 7 (2000), pp. 270–88, 275, and Stack, *Nature and Artifice*, ch. 5. For more on George's critique of Malthus, see J. Horner, 'Henry George on Thomas Robert Malthus: abundance vs. scarcity', *American Journal of Economics and Sociology*, 56 (1997), pp. 595–608.
7. Land nationalization was to prove a far more popular solution.
8. H. George, *Progress and Poverty* (London, 1953), p. 184.
9. Ibid., p. 55.
10. Ibid., p. 56.
11. Ibid., p. 184.
12. Ibid., pp. 186, 188.
13. Ibid., pp. 189–90, 194.
14. Ibid., p. 191.
15. Ibid., p. 193.
16. A. R. Wallace, *My Life: A record of events and opinions* (London, 1905), vol. 2, p. 14; Jones, *Henry George*, p. 68.
17. In 1881 Wallace became President of the Land Nationalisation Society. See M. Gaffney, 'Alfred Russel Wallace's campaign to nationalise land: how Darwin's peer learned from John Stuart Mill and became Henry George's ally', *American Journal of Economics and Sociology*, 56 (1997), pp. 609–19.
18. George, *Progress and Poverty*, pp. 215–16.
19. C. Tsuzuki, *The Life of Eleanor Marx: A socialist tragedy* (Oxford, 1967), pp. 77–8.
20. See J. Callaghan, *Socialism in Britain since 1884* (Oxford, 1990), p. 55.
21. See, for example, Lewis S. Feuer, 'Marxian tragedians: a death in the family', *Encounter* 5, XIX (1962), pp. 23–32.
22. E. Royle, *Radicals, Secularists and Republicans: Popular freethought in Britain, 1866–1915* (Manchester, 1980), p. 105. The best short account of Aveling's relationship with Darwin is in J. Moore, 'Freethought, secularism, agnosticism: the case of Charles Darwin', in G. Parsons (ed.), *Religion in Victorian Britain*, vol. I: *Traditions* (Manchester, 1988), pp. 274–319, especially pp. 309–12.

23. E. B. Aveling, *The Darwinian Theory: Its meaning, difficulties, evidence, history* (London, 1884), p. 48.
24. E. B. Aveling, *The Curse of Capital* (London, 1884), p. 163.
25. Ibid., p. 164.
26. Ibid., p. 168.
27. See A. Besant, 'Radicalism and socialism', in her *Essays in Socialism* (London, 1887).
28. Aveling, *Curse*, p. 168.
29. Aveling told his radical audience, 'You are an advance upon Liberalism; as Liberalism is an advance upon Whiggism, Whiggism on Conservatism, Conservatism on Toryism. And as men progress from the lower to the higher, the next step from Radicalism is Socialism. The difference, however, between the position of Radicalism and that of Socialism is much greater than between either of the other classes' (Aveling, *Curse*, p. 165).
30. E. B. Aveling, *A Godless Life the Happiest and Most Useful* (London, n.d.), p. 7; E. B. Aveling, *The Irreligion of Science* (London, 1881), p. 3. For the context of Aveling's remarks see J. W. Draper, *History of the Conflict between Religion and Science* (New York, 1974); O. Chadwick, *The Secularization of the European Mind in the Nineteenth Century* (Cambridge, 1975), ch. 7.
31. L. S. Feuer, 'Is the "Darwin–Marx Correspondence" authentic?', *Annals of Science*, 32 (1975), pp. 1–12; R. Colp Jr, 'The contacts of Charles Darwin with Edward Aveling and Karl Marx', *Annals of Science*, 33 (1976), pp. 387–94; M. A. Fay, 'Did Marx offer to dedicate *Capital* to Darwin? A reassessment of the evidence', *Journal of the History of Ideas*, 39 (1978), pp. 133–46; L. S. Feuer, 'The case of the Marx–Darwin letter', *Encounter* (October 1978), pp. 62–78.
32. E. B. Aveling, *The Religious Views of Charles Darwin* (London, 1883); *National Reformer: Radical Advocate and Freethought Journal*, 16 March 1884.
33. Royle, *Radicals, Secularists and Republicans*, pp. 171–2.
34. E. B. Aveling, *The Gospel of Evolution* (London, 1884), pp. 36, 48.
35. 'Mental evolution in animals', *National Reformer*, xliii (1884), pp. 210–11.
36. Aveling, *Gospel*, p. 39.
37. E. B. Aveling, *The Origin of Man* (London, 1884), p. 32.
38. 'Mind as a function of the nervous system', *National Reformer*, xxxix (1882), pp. 469–70; xl (1882), pp. 3–4; xl (1882), pp. 21–2.
39. Aveling, *Origin*, pp. 32, 48.
40. 'The commune of plants and animals', *National Reformer*, xlii (1883), pp. 371–2.
41. Aveling, *Origin*, p. 4.
42. Ibid., p. 39.
43. Ibid., p. 3.
44. 'Mind as a function of the nervous system', pp. 21–2.
45. See E. B. Aveling and E. Marx, *Shelley's Socialism: Two lectures* (Manchester, 1947).
46. 'To their mother, the wayward children are returning at last. Estranged from her for generations, all save a faithful few, wandering after strange gods, their feet are tending homewards to-day. Our heart's desire is that the kingdom of nature may come' (E. B. Aveling, *God dies: Nature remains*, London, 1881, p. 7).
47. P. Singer, *A Darwinian Left: Politics, evolution and co-operation* (London, 1999), pp. 19–20.
48. Todes has made a very convincing case for this, but is opposed by Kinna. D. P. Todes, 'Darwin's Malthusian metaphor and Russian evolutionary thought,

1859–1917', *Isis*, 78 (1987), pp. 537–51; R. Kinna, 'Kropotkin's theory of mutual aid in historical context', *International Review of Social History*, 40 (1995), pp. 259–83.

49. Todes, 'Darwin's Malthusian metaphor', p. 547.

50. Peter Kropotkin, *Mutual Aid: A factor in evolution*, with an introductory essay, 'Mutual aid and the social significance of Darwinism' (London, 1993), pp. 12–13.

51. See D. P. Todes, *Darwin without Malthus: The struggle for existence in Russian evolutionary thought* (Oxford, 1989), especially pp. 123–42.

52. Kinna, 'Kropotkin's theory', pp. 261–70.

53. T. H. Huxley, 'The struggle for existence: a programme', *The Nineteenth Century*, 23 (1888), pp. 161–80, was supplemented by his second Romanes lecture in May 1893, published as T. H. Huxley, *Evolution and Ethics* (London, 1893).

54. Kinna, 'Kropotkin's theory', p. 277.

55. Besides Todes, for the Russian evolutionary tradition, see J. A. Rogers, 'The reception of Darwin's *Origin of Species* by Russian scientists', *Isis*, 64 (1973), pp. 484–504; F. Sudro and M. Acanfora, 'Darwin and Russian evolutionary biology', in D. Kohn (ed.), *The Darwinian Heritage* (Princeton, 1985), pp. 731–49.

56. The chapters of *Mutual Aid* dealt in turn with animals, savages, the barbarous, medieval cities and contemporary society.

57. Kinna, 'Kropotkin's theory', p. 282.

58. 'Mutual aid among animals', xxviii (1890), pp. 337–54; 'Mutual aid among animals', xxviii (1890), pp. 699–719; 'Mutual aid among savages', xxix (1891), pp. 538–59; 'Mutual aid among the barbarians', xxxi (1891), pp. 101–22; 'Mutual aid in the medieval city', xxxvi (1894), pp. 183–202; 'Mutual aid in the medieval city', xxxvi (1894), pp. 397–418; 'Mutual aid amongst modern men', xxxix (1896), pp. 65–86; 'Mutual aid amongst ourselves', xxxix (1896), pp. 914–36.

59. See for example P. Kropotkin, 'The inheritance of acquired characters: theoretical difficulties', *Nineteenth Century*, lxxi (1912), pp. 511–31.

Chapter 4 'Social evolution is exasperatingly slow, isn't it sweetheart?'

1. '[In *The Iron Heel*] Jack London provides the reader with some insight into the icons, images, and language of a socialism mediated by a man who, probably more than any other writer of his generation, was profoundly and consciously affected by his times' (F. Shor, 'Power, gender and ideological discourse in "The Iron Heel"', in L. Cassuto and J. C. Reesman (eds), *Rereading Jack London* (Stanford, 1996), pp. 75–91, 91).

2. J. London, *The Iron Heel* (Chatham, 1996), p. 4.

3. Ibid., p. 100.

4. In a footnote Meredith explains, 'The people of the abyss – this phrase was struck out by the genius of H. G. Wells in the late nineteenth century AD' (ibid., p. 138fn). London used it in his 1903 study of the people of the East End of London: see J. London, *The People of the Abyss* (London, 1977).

5. See especially ch. 21 'The roaring abysmal beast' and ch. 22 'The Chicago Commune'. For Le Bon see J. S. McClelland, *The Crowd and the Mob: From*

Plato to Canetti (London, 1989); R. A. Nye, *The Origin of Crowd Psychology: Gustave Le Bon and the critics of mass democracy in the Third Republic* (London, 1975).

6. Disparaging references to 'our machine civilization' are scattered throughout *The Iron Heel*: see especially pp. 180, 177–8, 108, 122, 165, 170.

7. Ibid., pp. 54, 57, 38, 89, 104. In 'Goliah', a short story published soon after *The Iron Heel*, the leading character was a scientist rather than a proletarian, but retained the same socialist cast. See Shor, 'Power, gender and ideological discourse', p. 87.

8. Ibid., pp. 77, 125.

9. Ibid., pp. 71–3.

10. C. Johnston, *Jack London – An American Radical?* (London, 1984), pp. 43–5.

11. J. London, *Martin Eden* (Harmondsworth, 1909), pp. 94–5.

12. Quoted in J. Callaghan, *Socialism in Britain since 1884* (Oxford, 1980), p. 33.

13. R. M. Young, 'Herbert Spencer and inevitable progress', *History Today*, 37 (1987), p. 18.

14. No existing study of Spencer is wholly satisfactory, but perhaps this has as much to do with the subject as with his biographers. See J. D. Y. Peel, *Herbert Spencer: The evolution of a sociologist* (London, 1971); D. Wiltshire, *The Social and Political Thought of Herbert Spencer* (Oxford, 1978); T. S. Gray, *The Political Philosophy of Herbert Spencer: Individualism and organicism* (Aldershot, 1996).

15. See M. Taylor, *Men Versus the State* (Oxford, 1992), chs 2 and 3. According to Hawkins, Taylor underestimates the 'Darwinian' character of Spencer's thought (M. Hawkins, *Social Darwinism in European and American Thought 1860–1945: Nature as model and nature as threat* (Cambridgc, 1997), pp. 83fn, 87–8).

16. H. Spencer, *The Man Versus the State, with four essays on politics and society* (Harmondsworth, 1969), pp. 108–9; H. Spencer, *The Study of Sociology* (London, 1878), pp. 250, 253.

17. For Spencer's positive reinterpretation of Malthus see H. Spencer, 'A theory of population deduced from the general law of animal fertility', *Westminster Review*, 1 (1852), pp. 468–501.

18. Hawkins, *Social Darwinism*, pp. 98–103.

19. See, for example, the preface to H. Spencer, *The Inadequacy of Natural Selection* (London, 1893).

20. Ibid., p. 45.

21 Cf. Bowler, *Evolution: The history of an idea* (Berkeley, 1989), pp. 267, 272, who argues that Spencer sees competition merely as a stimulus to effort, with Hawkins, *Social Darwinism*, p. 88, who concludes that 'for Spencer, evolutionary progress entailed a continuous purging of the unfit; he had no vision of these latter improving their position and moving up the evolutionary ladder'.

22. Darwin, *Descent*, vol. I, p. 101.

23. H. Spencer, 'The social organism', in his *Essays: Scientific, political and speculative* (London, 1863), pp. 388–432, 392, 394.

24. Ibid., pp. 394–401. On Spencer's treatment of the 'social organism' see especially T. S. Gray, 'Herbert Spencer: individualist or organicist?', *Political Studies*, xxxiii (1985), pp. 236–53; W. M. Simon, 'Herbert Spencer and the social organism', *Journal of the History of Ideas*, 21 (1960), pp. 294–9.

25. L. Gronlund, *The Cooperative Commonwealth in its Outlines: An exposition of modern socialism* (Cambridge, Mass., 1965), p. 246.

26. Ibid., pp. 88, 80, 81.

27. W. C. Coleman, *Biology in the Nineteenth Century: Problems of form, function and transformation* (Cambridge, 1971), pp. 43–5.

28. Both, of course, were anti-Benthamite. Cf. H. Spencer, *Social Statics; or, The conditions essential to human happiness specified, and the first of them developed* (London, 1851) with T. Carlyle, 'Sign of the times', in his *Critical and Miscellaneous Essays*, vol. II (London, 1839), pp. 143–71.

29. Spencer, 'Social organism', p. 385.

30. T. H. Huxley, *Critiques and Addresses* (London, 1873), pp. 18–19.

31. A. Besant, *Why I am a Socialist* (London, 1886), pp. 2–3.

32. A. Besant, *The Evolution of Society* (London, 1886), p. 24.

33. Ibid.

34. Besant, *Why*, p. 3.

35. Peel, *Spencer*, ch. 8 'Militancy and industrialism', pp. 192–223.

36. Besant, *Evolution*, p. 4.

37. Ibid., pp. 15–16.

38. Ibid., p. 22.

39. Besant, *Why*, p. 3.

40. Ibid.

41. Robert Barltrop, for example, is wrong to conclude that 'The message is of Jack's loss of faith' (R. Barltrop, *Jack London, the Man, the Writer, the Rebel* (London, 1978), p. 127). Quite the opposite.

42. Trotsky's review of *The Iron Heel* is reproduced in S. M. Nuernberg, *The Critical Response to Jack London* (Westport, 1983), pp. 137–8.

43. J. London, 'Revolution' in his *Revolution: Stories and essays* (London, 1979), pp. 32–48, 32.

44. Ibid., p. 36.

45. Ibid., p. 48: 'The Revolution is here now. Stop it who can.'

Chapter 5 Ramsay MacDonald: Ideologist of Evolutionary Socialism

1. J. Ramsay MacDonald, 'Preface' to E. Ferri, *Socialism and Positive Science (Darwin–Spencer–Marx)* (London, 1905), pp. vii–viii.

2. A recent poll of Labour MPs saw MacDonald rated the worst of Labour's twelve leaders, *New Statesman*, 28 February 2000.

3. Mary Agnes Hamilton, *J. Ramsay MacDonald* (London, 1929), pp. 22, 121.

4. R. Barker, 'Socialism and Progressivism in the political thought of Ramsay MacDonald', in A. J. A. Morris (ed.), *Edwardian Radicalism 1900–1914: Some aspects of British radicalism* (London, 1974), pp. 114–30.

5. P. Thane, 'Labour and local politics: radicalism, democracy and social reform, 1880–1914', in E. Biagini and A. Reid (eds), *Currents of Radicalism: Popular radicalism, organised labour and party politics in Britain, 1880–1914* (Cambridge, 1991), pp. 244–70.

6. D. Marquand, *Ramsay MacDonald* (London, 1977).

7. L. MacNeil Weir, *The Tragedy of Ramsay MacDonald: A political biography* (London, 1938), p. 41.

8. J. Ramsay MacDonald, *Socialism and Society* (London, 1905), pp. 88–9.

9. Marquand, *Ramsay MacDonald*, pp. 12, 21–2, 53; Lord Elton, *The Life of James Ramsay MacDonald (1866–1919)* (London, 1939), pp. 48–9, 56–7.
10. PRO 30/69, 779.
11. PRO 30/69, 794.
12. PRO 30/69, 1829.
13. Elton, *Life*, pp. 225–6.
14. *Socialist Review*, 7 (March–August 1911), p. 412.
15. *Socialist Review*, 5 (March–August 1910), p. 8.
16. A typical example is Jon Callaghan's comment that 'MacDonald's socialism bore as much resemblance to science as alchemy' (J. Callaghan, *Socialism in Britain since 1884* (Oxford, 1990), p. 68).This type of remark forecloses any serious analysis of the role that science played in the development of MacDonald's thought.
17. MacDonald, 'Preface' to E. Ferri, *Socialism and Positive Science*, pp. vii–viii.
18. Ibid.
19. J. R. MacDonald, *Socialism* (London, 1907), p. 16; J. R. MacDonald, *The Socialist Movement* (London, 1909), p. 15.
20. MacDonald, *Socialism*, p. 48; MacDonald, *Socialist Movement*, pp. 89, 246.
21. MacDonald, *Socialism and Government*, vol. I, p. 87.
22. Ibid., p.19.
23. MacDonald, *Socialism and Society*, pp. 98–9.
24. *Socialist Review*, 1 (March–August 1908), pp. 16–17.
25. J. R. MacDonald, 'A plea for Puritanism', *Socialist Review*, 8 (September 1911–February 1912), pp. 422–30.
26. MacDonald, *Socialism and Society*, p. 36.
27. 'Revolution is the end, not the means to the end' (MacDonald, *Socialist Movement*, pp. 115–16).
28. Weir, *Tragedy*, p. xi. See also S. Macintyre, *A Proletarian Science: Marxism in Britain 1917–1933* (Cambridge, 1980); G. Stedman Jones, 'Working-class culture', p. 238, in his *Languages of Class: Studies in English working-class history, 1832–1982* (Cambridge, 1983).
29. MacDonald, *Socialism and Society*, pp. 98–9, 36.
30. A. R. Wallace to J. Ramsay MacDonald, 26 January 1909, PRO, MDP, 30/69, 1153, fos 150–1.
31. See MacDonald, *Socialism and Government*, vol. I, p. 6.
32. J. R. MacDonald, *The New Charter: A programme of working class politics* (Dover, 1892).
33. Unless otherwise stated, all quotes in this section are from J. R. MacDonald, 'The influence of sex on social progress', PRO 30/69, 1061.
34. Thane, 'Labour and local politics', p. 269.
35. For details of Rainbow Circle discussions, see M. Freeden (ed.), *Minutes of the Rainbow Circle, 1894–1924* (London, 1989). See also M. Freeden, *The New Liberalism: An ideology of social reform* (Oxford, 1986), p. 149.
36. See, especially, S. den Otter, *British Idealism and Social Explanation: A study in late-Victorian social thought* (Oxford, 1996); A. Vincent and R. Plant, *Philosophy, Politics, and Citizenship: Life and thought of the British Idealists* (Oxford, 1984); P. J. Nicholson, *The Political Philosophy of the British Idealists* (Cambridge, 1990).
37. R. G. Collingwood, *An Autobiography* (Oxford, 1939), p. 16. A judgement

shared by others: see I.W. Burrow, *Evolution and Society: A study in Victorian social theory* (Cambridge, 1966), pp. 260–1.

38. M. Carter, 'Ball, Bosanquet and the legacy of T. H. Green', *History of Political Thought*, 20 (1999), pp. 674–94.

39. S. Collini, *Liberalism and Sociology* (Cambridge, 1983), p. 150.

40. D. G. Ritchie, 'Natural selection and the history of institutions', in his *Darwinism and Politics* (London, 1891), pp. 122, 141; D. G. Ritchie, 'Social evolution', in his *Studies in Political and Social Ethics* (London, 1902), pp. 1–2.

41. *Manchester Guardian*, 27 December 1909. Cf. MacDonald's review of Hobhouse's *What is Liberalism?* in *Socialist Review*, 8 (1911–12), pp. 46–53.

42. D. G. Ritchie, 'Darwin and Hegel', in his *Darwin and Hegel, with Other Philosophical Studies* (London, 1893), pp. 64–6.

43. den Otter, *British Idealism*, pp. 105–7.

44. MacDonald, *Socialism and Society*, pp. 92, 108.

45. Ibid., p. 102.

Chapter 6 Marx and Engels

1. P. Singer, *A Darwinian Left: Politics, evolution and cooperation* (London, 1999), p. 5.

2. Ibid., p. 4.

3. Ibid., p. 5, emphasis added.

4. Ibid.

5. Ibid., pp. 26, 21.

6. See, for example, A. Schmidt, *The Concept of Nature in Marx* (London, 1971).

7. See, for example, P. Marshall, *Nature's Web: An exploration of ecological thinking* (London, 1992).

8. Singer, *Darwinian Left*, p. 22. On Lysenko, see D. Jorasky, *The Lysenko Affair* (Cambridge, Mass., 1970).

9. M. Foucault, *The Order of Things: An archaeology of the human sciences* (London, 1970), p. 262.

10. K. Marx, 'Theses on Feuerbach', in K. Marx and F. Engels, *The German Ideology* (London, 1965), pp. 645–53.

11. P. Heyer, *Nature, Human Nature, and Society: Marx, Darwin, biology, and the human sciences* (London, 1982), p. 69.

12. The term was taken from Feuerbach.

13. For an excellent discussion of this topic see N. Geras, *Marx and Human Nature* (London, 1983); K. Marx and F. Engels, *The Holy Family, or Critique of Critical Critique* (Moscow, 1956).

14. F. Engels, *The Origin of the Family, Private Property, and the State* (Harmondsworth, 1985), p. 146. For a critique of this strand in Marxism, see S. Timparano, *On Materialism* (London, 1980).

15. A. R. Wallace, *The Wonderful Century: Its successes and its failures* (London, 1898).

16. F. Engels, *Dialectics of Nature* (London, 1955), p. 242.

17. Ibid., p. 241.

18. Ibid., pp. 242–3.

19. K. Marx, *Early Writings* (Harmondsworth, 1984), p. 398.

20. Ibid., p. 348.

21. F. Engels, *Anti-Dühring: Herr Eugen Dühring's Revolution in Science* (London, 1975), p. 15.

22. Heyer, *Nature, Human Nature, and Society*, p. 14.

23. Ibid., p. 12.

24. W. Liebknecht, *Karl Marx: Biographical memoirs* (London, 1972), pp. 52–3, 64.

25. Engels, *Anti-Dühring*, p. 11.

26. Liebknecht, *Karl Marx*, pp. 91–2.

27. Marx to Engels, 19 December 1860, in K. Marx and F. Engels, *Collected Works*, vol. 41 (London, 1985), p. 232. See also Marx to F. Lasalle, December 1860, ibid., p. 246.

28. Marx to Engels, 7 August 1866, *Collected Works*, vol. 42 (London, 1987), pp. 304–5. See also Marx to Engels, 3 October 1866, ibid., p. 322; and Engels to Marx, 5 October 1866, ibid., p. 323.

29. R. Colp, Jr, 'The myth of the Darwin–Marx letter', *History of Political Economy*, 14 (1982), p. 462.

30. K. Marx, *Capital: A critique of political economy* (London, 1990), vol. 1, pp. 461fn, 493–4fn.

31. G. Lichtheim, *A Short History of Socialism* (London, 1971), pp. 88, 187.

32. See Engels to Marx, 30 May 1873, reproduced in W. O. Henderson (ed.), *Engels: Selected writings* (Harmondsworth, 1967), pp. 393–5; Engels, *Anti-Dühring*, p. 15.

33. Engels, *Anti-Dühring*, p. 42.

34. Ibid., pp. 9–10.

35. Schmidt, *Nature in Marx*, pp. 51, 57.

36. See, for example, S. Avineri, *The Social and Political Thought of Karl Marx* (Cambridge, 1969), p. 70.

37. See J. B. S. Haldane, *The Marxist Philosophy of the Sciences* (New York, 1969).

38. Singer, *Darwinian Left*, pp. 22, 26.

39. R. L. Meek, *Marx and Engels on Malthus: Selections from the writings of Marx and Engels dealing with the views of Thomas Robert Malthus* (London, 1953); F. Engels, 'Outline of a Critique of Political Economy', in Henderson (ed.), *Engels: Selected writings*, p. 150.

40. Marx, *Capital*, p. 461fn; F. Engels, *Socialism: Utopian and scientific* (Moscow, 1978), p. 53.

41. Engels, *Anti-Dühring*, p. 86.

42. K. Marx, *The Poverty of Philosophy* (Moscow, 1957), p. 146.

43. Engels, *Dialectics*, pp. 39–40.

44. Singer, *Darwinian Left*, p. 21.

45. F. Engels, 'The part played by labour in the transition from ape to man', in Engels, *Dialectics*.

46. Engels, *Anti-Dühring*, p. 18.

47. F. Engels, 'Natural science in the spirit world', in Engels, *Dialectics*, p. 80.

48. Engels recalled attending a lecture in Manchester in the winter of 1843–4 by the phrenologist Spencer Hall. Hall put volunteers from the audience into a magnetic sleep and then demonstrated the veracity of phrenology by inducing behaviour in accordance with whichever phrenological organ he touched. The lecture culminated with Hall touching the organ of veneration on a female volunteer, who immediately fell to her knees, hands in prayer, depicting an angel, and none too subtly proving the existence of God. Engels was unimpressed. He and a

friend hurried home to practise replicating Hall's performance with a twelve-year-old boy whom they managed to mesmerize. Engels' encounter with this 'very mediocre charlatan' would have remained forgotten had Wallace not cited Hall's lectures as an authority in favour of mesmerism in his *On Miracles and Modern Spiritualism* (1875). Ibid., p. 69.

49. Ibid.

50. Engels, *Anti-Dühring*, p. 31.

51. Engels, *Dialectics*, p. 54.

52. K. Marx, *Early Writings* (Harmondsworth, 1984), p. 328.

53. Ibid.

54. Heyer, *Nature, Human Nature, and Society*, pp. 58–61.

55. Engels, *Dialectics*, p. 402.

56. C. Darwin, *The Origin of Species by means of natural selection or the preservation of favoured races in the struggle for life* (Harmondsworth, 1985), p. 458.

57. Engels, *Dialectics*, p. 49. Marx agreed. Finding competition in nature, social Darwinism 'conversely considers this a conclusive reason for human society never to emancipate itself from its bestiality' (Marx to Paul and Laura Lafargue, Marx and Engels, *Collected Works*, vol. 43 (London, 1988), p. 217).

58. F. Engels, 'Ludwig Feuerbach and the end of Classical German Philosophy', in K. Marx and F. Engels, *Collected Works*, vol. 26 (London, 1990), p. 379.

59. Marx, *Early Writings*, p. 353; 'Natural science will in time subsume the science of man just as the science of man will subsume natural science; there will be *one* science' (ibid., p. 355).

60. Ibid., p. 391.

61. Engels, *Origin of the Family*, p. 64; Engels, *Dialectics*, pp. 402–5.

62. Marx to Engels, 14 November 1868. On Lange, Marx to Kugelmann, 27 June 1870.

63. He even quoted Vico: 'human history differs from natural history in that we have made the former' (Marx, *Capital*, pp. 493–4fn).

64. Engels, *Anti-Dühring*, p. 33.

65. Marx, *Capital*, pp. 283–6; Engels, *Dialectics*, pp. 230, 240, *passim.*

66. Engels, *Socialism: Utopian and scientific*, p. 9.

67. Ibid., p. 39, *passim.*

Chapter 7 The Revisionist Controversy

1. See H. Tudor and D. M. Tudor (eds), *Marxism and Social Democracy: The revisionist debate 1896–1898* (Cambridge, 1988). For the evolutionary angle, see R. Weikart, *Socialist Darwinism: Evolution in German socialist thought from Marx to Bernstein* (London, 1999).

2. R. Miliband, *Marxism and Politics* (Oxford, 1988), p. 121.

3. *Socialist Review*, 6 (September 1910–February 1911), p. 402; *Socialist Review*, 1 (March–August 1908), pp. 16–17.

4. M. B. Steger, *The Quest for Evolutionary Socialism: Eduard Bernstein and social democracy* (Cambridge, 1997), p. 47.

5. Ibid., pp. 66–9.

6. E. Bernstein, *Evolutionary Socialism: A criticism and affirmation* (London, 1909), pp. 213–14.

7. Ibid., pp. 102, 209–10.

8. Ibid., p. 210.

9. Ibid., pp. x, xii, xiii.

10. '[I]f I may be allowed to use an image of Lasalle – that it is Marx finally who carries the point against Marx' (ibid., p. 27).

11. Ibid., pp. 200–1.

12. Ibid., p. 197.

13. Ibid., p. 206.

14. E. Bernstein to J. Ramsay MacDonald, 1909. PRO 30/69, 1153, f. 20. See also P. Gay, *The Dilemma of Democratic Socialism: Eduard Bernstein's challenge to Marx* (New York, 1962), pp. 87–8, 146–7. Bernstein wrote a glowing review of MacDonald's *Socialism and Society* and provided the foreword to the German edition of *Socialism and Government*. See Weikart, *Socialist Darwinism*, p. 204.

15. M. Hawkins, *Social Darwinism in European and American Thought 1860–1945: Nature as model and nature as threat* (Cambridge, 1997), p. 158fn.

16. Bernstein, *Evolutionary Socialism*, p. 49.

17. Ibid., pp. 103, 105.

18. Ibid., p. 51.

19. Ibid., p. xxii.

20. Ibid.

21. Ibid., p. 148.

22. L. Trotsky, *Political Profiles* (London, 1972), p. 69.

23. P. Goode (ed.), *Karl Kautsky: Selected political writings* (London, 1983), p. 15.

24. Steger, *Bernstein*, p. 47. For a detailed consideration of Kautsky's thought, see G. P. Steenson, *Karl Kautsky 1854–1938: Marxism in the classical years* (Pittsburgh, 1979); M. Salvadori, *Karl Kautsky and the Socialist Revolution 1880–1938* (London, 1979). See also Weikart, *Socialist Darwinism*, chapter 6, and C. Rafferty, 'Karl Kautsky – between Darwin and Marx', *Australian Journal of Politics and History*, 38 (1990).

25. See P. Nettl, 'The German Social Democratic Party, as a political model', *Past and Present*, 30 (1965), p. 73.

26. 'Kautsky on More's Utopia' in S. Mukherje and S. Ramaswamy (eds), *World's Greatest Socialist Thinkers – Thomas More (1478–1535)* (New Delhi, 1998), p. 276.

27. K. Kautsky, 'Leben, Wissenschaft und Ethik', *Die Neue Zeit*, xxiv (1906), pp. 516–29, in Goode (ed.), *Kautsky*, p. 48.

28. See T. Bottomore and P. Goode (eds), *Austro-Marxism* (Oxford, 1978), pp. 78–84.

29. K. Kautsky, *Ethik und materialistische Geschichtsauffassung* (Stuttgart, 1906), in Goode (ed.), *Kautsky*, pp. 33–45, 38.

30. Ibid., p. 42.

31. Ibid., p. 43.

32. Ibid., pp. 42–3.

33. K. Kautsky, *The Road to Power: Political reflections on growing into the revolution* (New Jersey, 1996), p. 17.

34. 'Thus growing into socialism means growing into great struggles that will convulse the entire political system, that must continually become more powerful and can end only with the defeat and expropriation of the capitalist class' (ibid., p. 18).

35. Ibid., pp. 9, 91.

36. Ibid., p. 18.

37. Goode (ed.), *Kautsky*, pp. 29–30.

38. Kautsky, *Road*, pp. 33–41.
39. Ibid., p. 3.
40. Goode (ed.), *Kautsky*, p. 30.
41. Kautsky, *Road*, p. 23.
42. Ibid.
43. Ibid., pp. 23–4.
44. 'The capitalists' will to live meets with conditions that drive them to bend the workers' will and make it subservient to themselves. Without this bending of the will, there would be no capitalist profits; no capitalists could exist. On the other hand, the workers' will to live drives these to rebel against the capitalists' will. Hence the class struggle' (ibid., p. 24).
45. H. Menai, 'Socialism and the organic conception of society', *Socialist Review,* 5 (March 1910–August 1910), p. 398.
46. K. Kautsky, *The Dictatorship of the Proletariat* (London, 1919), p. 55.
47. Ibid., p. 103.
48. Ibid., p. 5.
49. Ibid., p. 7.
50. Ibid., pp. 98–9.

Chapter 8 Eugenics and Parasitology

1. See D. J. Kelves, *In the Name of Eugenics: Genetics and the uses of human heredity* (Harmondsworth, 1992).
2. C. Darwin, *The Descent of Man, and selection in relation to sex* (London, 1871), vol. I, pp. 168–9.
3. C. Darwin, *The Origin of Species by means of natural selection, or the preservation of favoured races in the struggle for life* (London, 1985), pp. 98–9.
4. K. Pearson, 'Socialism and natural selection', *The Fortnightly Review*, cccxxxi (1894), pp. 1–21.
5. F. Galton, *Memories of My Life* (London, 1908), p. 321; D. W. Forrest, *Francis Galton: The life and work of a Victorian genius* (London, 1974), p. 260.
6. See W. H. Schneider, *Quality and Quantity: The quest for biological regeneration in twentieth-century France* (Cambridge, 1990).
7. See W. H. Schneider, 'Towards the improvement of the human race: the history of eugenics in France', *Journal of Modern History*, 54 (1982), pp. 268–91; M. Freeden, 'Eugenics and progressive thought: a study in ideological affinity', *Historical Journal*, 22 (1979), pp. 645–71.
8. That was the 1859 edition of the *Origin*. The 1868 edition softened the line to 'the laws of inheritance are for the most part unknown'. See W. George, *Darwin* (London, 1982), pp. 59–62.
9. K. Pearson, *Nature and Nurture: The problem of the future* (London, 1910), p. 27.
10. G. Claeys' paper at the 'Long Eighteenth Century Seminar', Institute of Historical Research, London, 21 March 2001; G. Claeys (ed.), *Modern British Utopias*, 8 vols (London, 1997).
11. See Galton Papers, Galton Archive, University College London.
12. C. Shaw, 'Eliminating the yahoo: eugenics, social darwinism and five Fabians', *History of Political Thought*, 8 (1987), pp. 521–44. See I. Britain, *Fabianism and Culture* (Cambridge, 1982). A good example of the Fabian embrace of eugenics can be found in Bernard Shaw's 'Sixty years of Fabianism essay', in G. B. Shaw (ed.), *Fabian Essays*, Jubilee Edition (London, 1948), pp. 207–31.

13. Darwin, *Origin*, pp. 186–7.

14. J. D. Young, 'Militancy, English socialism and the *Ragged Trousered Philanthropists*', *Journal of Contemporary History*, 20 (1985), pp. 283–303, 290–1.

15. A. Besant, *Evolution of Society* (London, 1886), pp. 20–1.

16. A. Besant, *Why I am a Socialist* (London, 1886), p. 8. Besant toyed with eugenic notions: 'There can be no doubt that a human aristocracy might be bred, by making men and women who showed in a marked degree the qualities which might be selected as admirable' (Besant, *Evolution*, p. 8, although cf. ibid., pp. 13–14).

17. E. R. Lankester, *Degeneration: A chapter in Darwinism* (London, 1880), p. 29.

18. On degeneration see, especially, D. Pick, *Faces of Degeneration: A European disorder, c.1848–c.1918* (Cambridge, 1989).

19. L. Gronlund, *The Cooperative Commonwealth* (Cambridge, Mass., 1965), pp. 44–5.

20. 'Our parasites are useless workers; our Vampires are not better than thieves and swindlers' (ibid., p. 51).

21. Ibid., p. 94.

22. Ibid., p. 52.

23. J. Polasky, *The Democratic Socialism of Emile Vandervelde: Between reform and revolution* (Washington, 1995), pp. 12–13.

24. P. Geddes, 'Preface' to J. Massart and E. Vandervelde, *Parasitism: Organic and social* (London, 1895), p. v.

25. Polasky, *Vandervelde*, p. 12.

26. Massart and Vandervelde, *Parasitism*, p. 1.

27. Ibid., p. 2.

28. Ibid., pp. 121, 100.

29. Ibid., pp. 18–22.

30. Ibid., pp. 71–2.

31. Ibid., p. 83.

32. Ibid., pp. 71–2.

33. Ibid., pp. 121–2.

34. Polasky, *Vandervelde*, p. 263.

35. E. Vandervelde, *Collectivism and Industrial Revolution* (London, 1907), p. 12.

36. Ibid., pp. 41, 47, 206.

37. Ibid., p. 82.

38. Massart and Vandervelde, *Parasitism*, p. 22.

39. Freeden, 'Eugenics and progressive thought', p. 653.

40. See R. First and A. Scott (eds), *Olive Schreiner* (London, 1980).

41. O. Schreiner, *Woman and Labour* (London, 1978), p. 108.

42. S. Ledger, *The New Woman, Fiction and Feminism at the Fin de Siècle* (London, 1998), p. 75.

43. S. Tasker, 'Civilisation and its Parasites, an enquiry into the racial science and utopian aspirations of Olive Schreiner in relation to Francis Galton's eugenics, and the discourses of degeneration in the late Victorian and Edwardian period', MA dissertation, Queen Mary College, University of London, 1998.

44. Schreiner, *Woman*, p. 27.

45. Darwin, *Descent*, vol. II, pp. 326–9.

46. R. Brandon, *The New Women and the Old Men: Love, sex and the woman question* (London, 1992), pp. 44–94, 54–5.

47. Schreiner, *Woman*, p. 108.
48. Ibid., pp. 52, 100, 66.
49. Ibid., pp. 129–30.
50. Ibid., p. 101.
51. Ibid., pp. 12–13; R. Rive (ed.), *Olive Schreiner, Letters, Vol. 1 1871–1899* (Oxford, 1988), p. 43.

Chapter 9 The ILP and the Socialist Library

1. 'The realisation of Socialism is not near. Society in these later days does not grow by revolution and sudden change, but by the slow enlightenment of the democratic intelligence and conscience' (Anon. [J. R. MacDonald], *The ILP: What it is, and where it stands* (London, n.d.), p. 3).
2. ILP, *Report of the Sixth Annual Conference, April 1898* (London, 1898), p. 27.
3. *Socialist Review*, 1 (March–August 1908), pp. 16–17.
4. E. P. Thompson, 'Homage to Tom Maguire', in A. Briggs and J. Saville (eds), *Essays in Labour History* (London, 1960), pp. 276–316, planted a seed that has borne considerable fruit. See especially J. A. Jowitt and R. H. S. Taylor (eds), *Bradford 1890–1914: The cradle of the Independent Labour Party* (Leeds, 1980); D. Clark, *Colne Valley: Radicalism to socialism* (London, 1981); K. Laybourn and D. James (eds), *'The Rising Sun of Socialism': The ILP in the textile districts of the West Riding of Yorkshire between 1890 and 1914 – a collection of essays* (Leeds, 1991); A. McKinlay and R. J. Morris (eds), *The ILP on Clydeside, 1893–1932, from foundation to disintegration* (Manchester, 1991).
5. R. E. Dowse, *Left in the Centre: The Independent Labour Party 1893–1940* (London, 1966), dispenses with the pre-First World War period in twenty pages. D. Howell, *British Workers and the Independent Labour Party, 1888–1906* (Manchester, 1984), is better, but obscures the overall ideological picture in a mass of local detail. See also D. Tanner, *Political Change and the Labour Party* (Cambridge, 1990).
6. The seminal article on ethical socialism is S. Yeo, 'A New Life: the religion of socialism in Britain, 1883–1896', *History Workshop Journal*, 4 (1977), pp. 5–56. See also L. Smith, 'Religion and the ILP', in D. James, T. Jowitt and K. Laybourn (eds), *The Centennial History of the Independent Labour Party* (Halifax, 1992), pp. 299–310.
7. So strong is the tendency to see the British labour movement as ideologically adrift from the continent that even those inclined to narrate the history of the British left within a European framework see Britain as distinct until 1918. See D. Sassoon, *One Hundred Years of Socialism: The west European left in the twentieth century* (London, 1996), p. 16.
8. This is not recognised by Howell, *British Workers*, pp. 5, 10, and his negative assessment is echoed in B. Winter, *The ILP: Past and present* (Leeds, 1993), p. 11, and M. Crick, '"A call to arms": the struggle for socialist unity in Britain, 1883–1914', in James, Jowitt and Laybourn (eds), *Centennial History*, pp. 181–204.
9. K. Hardie, *The Keir Hardie Calendar: A quotation from the writings of J. Keir Hardie for every day in the year* (Manchester, n.d. [1916?]), entry for 8 October.
10. See C. Steedman, *Childhood, Culture and Class in Britain: Margaret McMillan, 1860–1931* (London, 1990), pp. 156–72, 179.
11. McMillan lived in Bromley, Kent, and Kropotkin in nearby Chiselhurst.

12. See K. Laybourn, *Philip Snowden: A biography, 1864–1937* (Aldershot, 1988), pp. 10, 37. More generally see J. Laurent, 'Science, society and politics in late-nineteenth century England: a further look at Mechanics' Institutes', *Social Studies in Science*, 14 (1984), pp. 585–619.

13. L. Small, *Darwinism and Socialism* (London, 1908), p. 3.

14. Ibid., p. 4.

15. Ibid., p. 11.

16. Ibid., p. 13.

17. 'The Socialist Library Prospectus', reprinted in E. Ferri, *Socialism and Positive Science (Darwin–Spencer–Marx)* (London, 1905), pp. 175–8.

18. J. R. MacDonald, 'Preface' to J. Jaurès, *Studies in Socialism* (London, 1906), p. xiv; J. R. MacDonald, *Socialism and Society* (London, 1905), 2nd edn, pp. xi–xii.

19. G. B. Woolven, *Publications of the Independent Labour Party, 1893–1932* (Warwick, 1977), p. vii.

20. ILP, *Report of the Twelfth Annual Conference, April 1904* (London, 1904), p. 18.

21. ILP, *Report of the Thirteenth Annual Conference, April 1905* (London, 1905), pp. 18–19.

22. ILP, *Report of the Fifteenth Annual Conference, April 1907* (London, 1907), p. 12.

23. J. Bruce Glasier to J. Ramsay MacDonald, 12 March 1908. PRO, MDP, 30/69, 1152, fo. 117.

24. As MacDonald explained in his chairman's address to the 1907 conference. See ILP, *Report of the Fifteenth Annual Conference, April 1907* (London, 1907), p. 33.

25. Ibid., pp. 12, 10.

26. 'An Active Member', *The Revolution in the Baltic Provinces of Russia: A brief account of the activity of the Lettish Social Democratic Workers Party* (London, 1907).

27. ILP, *Report of the Eighteenth Annual Conference, April 1910* (London, 1910), pp. 13, 42–3.

28. ILP, *Report of the Twenty-first Annual Conference, April 1913* (London, 1913), p. 11.

29. The two series had different publication arrangements, and 'The ILP Library' was projected as an ongoing venture. See ILP, *Report of the Annual Conference, April 1920* (London, 1920), p. 21.

30. Hayward and Langdon-Davies made an educationalist plea for a free press, and appealed to Benjamin Kidd for scientific authority. See F. H. Hayward and B. N. Langdon-Davies, *Democracy and the Press* (Manchester, 1919), pp. vii, 69. The anonymous 'Editor's Preface' to the English translation of Kautsky reads as if written by MacDonald. See 'Editor's preface to the English edition', in K. Kautsky, *The Dictatorship of the Proletariat* (Manchester, 1919), pp. iii–vii. MacDonald had long been interested in translating Kautsky's writings. [Unknown] to J. R. MacDonald, PRO, MDP, 30/69, 1151, fo. 5.

31. MacDonald, 'Preface', p. v. On Lombroso, see D. Pick, *Faces of Degeneration: A European disorder, c.1848–c.1918* (Cambridge, 1989).

32. Ferri, *Socialism and Positive Science*, p. xi.

33. Ibid., p. 5.

34. Ibid., pp. 8, 8–11fn.

35. Ibid., p. 28.

36. Ibid., pp. 38, 42.
37. Ibid., p. 113.
38. Ibid., p. 92.
39. Ibid., p. 57.
40. Ibid., p. 114.
41. Ibid., p. 78.
42. Ibid., p. xii.
43. Ibid., p. 123.
44. Ibid., p. 135.
45. Ibid., p. 144.
46. Small, *Socialism and Darwinism*, p. 13.
47. Ibid.
48. S. Herbert, 'Socialism and the new science', *Socialist Review*, 1 (March–August 1908), pp. 305–9; F. C. Constable, 'Socialism and the new science: another view', *Socialist Review*, 2 (March–August 1908), pp. 598–608; S. Herbert, 'Socialism and the new science: a reply', *Socialist Review*, 1 (March–August 1908), pp. 792–800; M. Eden Paul, 'Socialism and science', *Socialist Review*, 3 (March–August 1909), pp. 95–106; F. H. Minett, 'Socialism and sham science', *Socialist Review*, 3 (March–August 1909), pp. 292–8.
49. F. C. Constable, 'Socialism and the survival of the fittest', *Socialist Review*, 5 (March–August 1910), pp. 254–60; H. Feblinger, 'Natural selection and non-selective elimination', *Socialist Review*, 8 (September 1911–February 1912), pp. 57–66; F. C. Constable, 'Eugenics and socialism', *Socialist Review*, 10 (September 1912–February 1913), pp. 17–25; Lancelot T. Hogben, 'Modern heredity and social science', *Socialist Review*, 16 (January–March 1919), pp. 147–56.
50. *Socialist Review*, 7 (March–August 1911), pp. 172, 412.
51. S. Herbert, 'Socialism and the new science: a reply', *Socialist Review*, 1 (March–August 1908), p. 795.
52. McMillan, *Child and the State*, pp. 73, 79, 81, 8, xii–xiii.
53. P. Snowden, *Socialism and the Drink Question* (London, 1908), pp. 3, 171–82, 7, 67.
54. J. R. MacDonald, The *Socialist Movement* (London, 1909), vol. II, pp. 4, 85–6, 95–9; vol. I, pp. 14–15.
55. *Socialist Review*, 7 (March 1911–August 1911), pp. 172, 412.
56. MacDonald, *Socialist Movement*, vol. I, p. 244.
57. J. R. MacDonald, *Socialism and Government* (London, 1909), vol. I, p. 21.
58. Ibid., vol. I, pp. 85–6.
59. F. Lee, *Fabianism and Colonialism: The life and thought of Lord Sydney Olivier* (London, 1988), pp. 223, 190.
60. Ibid., pp. 111, 29.
61. Ibid., pp. 189–90; S. Olivier, *White Capital and Coloured Labour* (London, 1906), p. 11.
62. Lee, *Fabianism and Colonialism*, p. 220; S. Olivier, 'Long views and short on black and white', *The Contemporary Review*, xx (1906), pp. 491–504.
63. C. Hall, 'The economy of intellectual prestige: Thomas Carlyle, John Stuart Mill and the case of Governor Eyre', *Cultural Critique*, 12 (1989), pp. 167–79.
64. Olivier, *White Capital*, p. 137.
65. Ibid., p. 2.
66. Ibid., p. 11.
67. Ibid., pp. 139, 147.
68. Ibid., p. 162.

69. M. N. El-Amin, 'Sydney Olivier on socialism and the colonies', *Review of Politics*, 39 (1977), pp. 521–39.
70. Olivier, *White Capital*, pp. 17, 12.
71. Ibid., p. 13.
72. Ibid., p.12.
73. Ibid., p. 59.
74. Ibid., p. 18; Lee, *Colonialism and Fabianism*, p. 113.
75. Olivier, *White Capital*, p. 29.
76. Ibid., p. 36.
77. Ibid., pp. 37–8.
78. Ibid., pp. 23–4.

Chapter 10 Neither Liberalism nor Marxism

1. Laura Vandervelde to J. R. MacDonald, 9 April 1904, PRO 30/69, 1148, ff. 32–3; E. Bernstein to J. R. MacDonald, 17 April 1908, PRO 30/69, 1152, ff. 44–9.
2. E. Vandervelde, *Collectivism and Industrial Evolution* (London, 1907), p. 12.
3. J. R. MacDonald, *Socialism and Government*, vol. 1 (London, 1909), pp. 3, 105–6; E. Bernstein, *Evolutionary Socialism: A criticism and affirmation* (London, 1909), p. 114; P. Snowden, *Socialism and the Drink Question* (London, 1908), pp. 44–5, *passim*.
4. M. McMillan, *The Child and the State* (London, 1911), p. 175.
5. R. McKibbin, 'Why was there no Marxism in Britain?', in his *The Ideologies of Class: Social relations in Britain, 1880–1950* (Oxford, 1991), pp. 1–41.
6. D. Marquand, *Ramsay MacDonald* (London, 1977), p. 113.
7. J. R. MacDonald, *Socialism To-day, being the chairman's address (expanded and amplified) delivered at the Synod Hall, Edinburgh, Easter 1909* (London, 1909).
8. J. R. MacDonald, *Syndicalism, a Critical Exposition* (London, 1912), pp. 52, 51, 6.
9. J. R. MacDonald, 'Syndicalism', *Socialist Review*, 8 (September 1911–February 1912), pp. 117–24.
10. Marquand, *MacDonald*, p. 257.
11. [Unknown] to J. R. MacDonald, PRO 30/69, 1121, f. 5.
12. J. R. MacDonald, *Parliament and Revolution* (London, 1919), pp. 11, 89, 39–40, 99–101, 93–4.
13. W. Newbold to J. R. MacDonald, 13 August 1919, PRO 30/69, 1163.
14. ILP, *Report of the Annual Conference, April 1919* (London, 1919), p. 21.
15. Ibid., p. 54.
16. ILP, *Report of the Annual Conference, April 1921* (London, 1921), p. 17.
17. M. A. Hamilton, *The Principles of Socialism* (London, 1921), p. 25.
18. J. R. MacDonald, *The History of the ILP* (London, 1921), p. 11.
19. K. Hardie and J. R. MacDonald, 'The Liberal collapse: the Independent Labour Party's programme', *The Nineteenth Century*, 45 (1899), pp. 20–38; 'Chairman's address', ILP, *Report of the Sixteenth Annual Conference, April 1908* (London, 1908), p. 39; MacDonald, *Socialism and Society*, p. 149.
20. MacDonald, *Socialism and Society*, pp. 152–3.
21. E. Biagini and A. Reid (eds), *Currents of Radicalism: Popular radicalism, organised labour and party politics in Britain, 1850–1914* (Cambridge, 1991).
22. J. R. MacDonald, *Socialism: Critical and constructive* (London, 1921),

pp. 48–55; J. R. MacDonald, *The Social Unrest: Its cause and solution* (London, 1913), pp. 9–28.

23. R. Gray, 'Class, politics and historical "revisionism"', *Social History*, 19 (1994), p. 218.

24. See, for example, *Socialist Review*, 1 (1908), pp. 15–16.

25. J. R. MacDonald, 'What is Liberalism? A review', *Socialist Review*, 8 (1911–12), pp. 46–53.

26. 'Socialism is a tendency, not a revealed dogma, and therefore it is modified in its forms of expression from generation to generation. The good remains the same, but the path twists and turns like every human path' (J. R. MacDonald, *The Socialist Movement* (London, 1909), p. 195).

27. J. Jaurès, *Studies in Socialism* (London, 1906), pp. 13–16.

Conclusion

1. R. Boyd, 'Metaphor and theory change: what is metaphor for?', in A. Ortony (ed.), *Metaphor and Thought* (Cambridge, 1979), pp. 356–408; A. Klamer and T. C. Leonard, 'So what's an economic metaphor?', in P. Mirowski (ed.), *Natural Images in Economic Thought* (Cambridge, 1994), pp. 20–49.

2. P. Mirowski, *More Heat than Light: Economics as Social Physics, Physics as Nature's Economics* (Cambridge, 1989).

3. J. Laurent, 'Science, society and politics in late nineteenth century England: a further look at Mechanics' Institutes', *Social Studies in Science*, 14 (1984), pp. 585–619.

4. See J. A. Paradis, *Thomas Huxley: Man's place in nature* (Lincoln, NE, 1978), p. 11 and *passim*; A. P. Barr, '"Common sense clarified": Thomas Henry Huxley's faith in truth', in A. P. Barr (ed.), *Thomas Henry Huxley's Place in Science and Letters* (London, 1997).

5. R. E. Dowse, *Left in the Centre: The Independent Labour Party 1893–1940* (London, 1966), pp. 122–3.

6. David Marquand, *Ramsay MacDonald* (London, 1977), p. 456.

7. Dowse, *Left in the* Centre, pp. 136, 78.

8. Ibid., p. 123.

9. W. Newbold to J. R. MacDonald, 13 August 1919, PRO 30/69, 1163. Keith Laybourn has identified a generational shift in the ILP. See K. Laybourn, *Philip Snowden: A biography, 1864–1937* (Aldershot, 1988), p. 135.

10. The party's 1928 policy document was, MacDonald noted approvingly, 'a synthesis between social theory and evolutionary action' (D. Watts, *Ramsay MacDonald: A Labour tragedy?* (Oxford, 1998), p. 131).

11. C. R. Attlee, *The Labour Party in Perspective* (London, 1937), pp. 33–61.

12. Ibid., pp. 59–60.

13. S. Olivier, *White Capital and Coloured Labour* (London, 1906), p. 58.

Index

ALSO FROM NEW CLARION PRESS

Genetic Politics: From eugenics to genome

Anne Kerr and Tom Shakespeare

Shortlisted for the *Sociology of Health and Illness Journal* 2003 book prize for outstanding contribution to medical sociology

> **'a thought-provoking book, laden with information and detailed historical records'** *Times Higher Education Supplement*
>
> **'[an] important new book . . . I would thoroughly recommend it for use in a wide range of educational settings, including the sociology of science and technology, medical sociology and disability studies.'** *Medical Sociology News*
>
> **'a very welcome and timely analysis of "the new genetics" from a disability rights perspective . . . an important text, offering a well-argued and reasoned analysis of current genetic policy'** *Disability and Society*
>
> **'a superb, historically rooted narrative . . . highly readable'** *Health, Risk and Society*

Genetic Politics explores the history of eugenics and the rise of contemporary genomics, identifying continuities and changes between the past and the present. Anne Kerr and Tom Shakespeare reject the two extreme positions that human genetics are either fatally corrupted by, or utterly immune from, eugenic influence. They argue that today's forms of genetic screening are far from equivalent to the eugenics of the past, but eugenics cannot simply be dismissed as bad science, or the product of totalitarian regimes, for its values and practices continue to shape genetics today.

Triumphalist accounts of scientific progress and the merits of individual choice mask how genetic technologies can undermine people's freedom, by intensifying genetic determinism and discrimination, individualizing responsibility for health and welfare, and stoking intolerance of diversity. Regulation is largely ineffectual at limiting these dangers because it is often guided by the goals of perfect health and commercial profit. The authors argue that we need to listen to the people directly affected by the new genetics technologies, especially disabled people and women, and to challenge the values and practices that shape genetics.

Anne Kerr is a lecturer in sociology at the University of York with specialist interests in genetics and gender. **Tom Shakespeare** is Director of Outreach at the Policy, Ethics and Life Sciences Research Institute, Newcastle, and has written widely on disability and genetics.

vii + 211 pages, illustrated
Published in 2002

Paperback	£12.95	$24.95	ISBN 1 873797 25 7
Hardback	£25.00	$49.95	ISBN 1 873797 26 5